Physical Chemistry in Action

Physical Chemistry in Action presents volumes which outline essential physicochemical principles and techniques needed for areas of interdisciplinary research. The scope and coverage includes all areas of research permeated by physical chemistry: organic and inorganic chemistry; biophysics, biochemistry and the life sciences; the pharmaceutical sciences; crystallography; materials sciences; and many more. This series is aimed at students, researchers and academics who require a fundamental knowledge of physical chemistry for working in their particular research field. The series publishes edited volumes, authored monographs and textbooks, and encourages contributions from field experts working in all of the various disciplines.

Emilia Wołowiec-Korecka

Carburising and Nitriding of Iron Alloys

Springer

Emilia Wołowiec-Korecka
Institute of Materials Science
and Engineering
Lodz University of Technology
Łódź, Poland

ISSN 2197-4349 ISSN 2197-4357 (electronic)
Physical Chemistry in Action
ISBN 978-3-031-59861-6 ISBN 978-3-031-59862-3 (eBook)
https://doi.org/10.1007/978-3-031-59862-3

This Springer imprint is published by the registered company Springer Nature Switzerland AG
The registered company address is: Gewerbestrasse 11, 6330 Cham, Switzerland

Paper in this product is recyclable.

I dedicate this book to my beloved husband, Maciej.

Preface

The actual conditions in which science operates are directing the research towards utilitarianism. Almost every application for research project funding these days includes a question about the advisability of the planned research and the viability of its implementation. An important question, therefore, is: how do we analyse and value the development trends of thermo-chemical treatment in order to correctly identify those that are prospective? Meanwhile, it is difficult to assess contemporary development trends, influenced by the uneven balance of the vast scientific literature and the scarcity of sources concerning practical applications. The aim of this book is to present a structured, albeit simplified, picture of modern thermo-chemical treatment to answer questions about the causes, relationships, and trends in this area of science. Cognition and explanation serve to understand the reality, and understanding is the ideal state: it allows you to optimise activities in the future without a widely conceived waste of time and resources. Tracing the development and attempting to bring together the scattered knowledge into a coherent study seems a worthwhile endeavour, although it is clear that it is impossible to provide a comprehensive and exhaustive description in a single volume.

Global thermal treatment and thermo-chemical treatment (TCT) remain fragmented, and the research and development efforts necessary to innovate the industry are only being pursued in certain segments, despite the fact that expectations as to the quality and precision in product manufacture are steadily increasing and the correct selection of treatment parameters and maximising their impact are becoming increasingly important. A key challenge of modern TCT is to achieve high repeatability and uniformity of manufactured products at high volume and mass scale.[1] Additional trends are set by the modern organisation of industrial production—observed especially in the automotive industry—where the concept of management by quality is becoming the standard for production management in companies. This new way of thinking and approaching management problems, using statistical process control,

[1] The global heat treating market size was valued at USD 100.73 billion in 2021 and is expected to expand at a compound annual growth rate of 3.4% from 2022 to 2030 (Grand View Research, 2021).

is the basis for an organisation to operate in a way where the quality of the manufactured product is the guiding element and at the same time the link between all spheres of activity.

The uncompromising market economics of the twenty-first century make the contemporary utilitarian development of thermo-chemical treatment possible in two ways: (1) by offering quality at a level unachievable for competitors or (2) by reducing manufacturing costs without losing quality. Both trends are evident in the current development of thermo-chemical treatment and are often linked in some way. However, finding their new areas of applicability is combined with developing knowledge of these processes. Today, many gaps in the contemporary state of knowledge are being filled based on high-performance computer calculations, which have become part of the canon of the modern scientific research workshop. Computational models and simulations contribute to both reducing manufacturing costs and improving production quality. For this reason, the following chapters discuss the contemporary development of thermo-chemical treatments against the background of the parallel development of computational capabilities, pointing out references in the literature, available scientific research and known practical applications.

The book is divided into seven chapters, discussing quenching, carburising and nitriding individually, although the issues sometimes overlap. Chapter 1 provides an introduction to the work and identifies the importance of materials engineering against the background of contemporary societal needs. Chapter 2 discusses thermo-chemical treatment and its development over the centuries. The modern challenges for thermo-chemical treatment globally are presented next. Chapters 3–5 present the three main development trends: reduction of thermal deformation in mass and high-volume production, TCT modelling and a broadly perceived reduction of the costs of conducting the treatment. These are identified in the literature at the dawn of the twenty-first century as important and necessary to be addressed quickly. This is the Author's subjective choice, as the list of important issues is much longer (this topic is discussed at length in Chap. 2). Chapter 6 describes the importance and contribution of computational methods to the development of thermo-chemical treatment. All the issues discussed and the challenges of the future are succinctly summarised in Chap. 7, which concludes the work.

This study provides a literature review of the development trends of modern thermo-chemical treatment and, in this respect, fills a gap in the literature on this topic. Due to the review nature of the work, it can serve as supplementary literature in subjects covering surface engineering for students of materials engineering and mechanical engineering.

Łódź, Poland — Emilia Wołowiec-Korecka
January 2024

Declarations

Competing Interests The author was an R&D manager of a research grant from the National Centre for Research and Development of Poland as part of project no. POIR.04.01.04-00-0087/15 entitled: "Equipment for high performance and precise heat treatment with a quenching deformation reduction system for direct application in downstream production chains of mechanical gearing and bearings". The research results presented on some figures and tables in Chaps. 3 and 4 were financed from this grant (Figs. 3.2, 4.2, 4.3, 4.4, 4.5, 4.6 and Table 3.7).

Ethics Approval None of the studies presented in this book were conducted on patients or animals.

Author's professional e-mail address: emilia.wolowiec-korecka@p.lodz.pl
Author's private e-mail address: e.wolowiec@gmail.com

Acknowledgements

This book would not have been created in its current form without the participation and support of others. Many scholars and friends helped me during the various stages of working on this book when I encountered various difficulties. It is not possible for me to list all the people who deserve it, but I would like to mention the most important ones:

I would like to thank Prof. Piotr Kula, who in many ways shaped my view of materials engineering, including the belief that there must be utilitarian aspects to scientific work.

I would like to thank Prof. Jerzy Michalski for his careful reading of the first version of the book and his accurate comments and suggestions, which helped to give it a much better final shape.

I would like to thank Prof. Leszek Klimek for answering my many questions about materials engineering and for being an inexhaustible source of literature.

I would like to thank Prof. Leszek A. Dobrzański for his kindness and for motivating me to "stay at the course of the work".

I would like to thank my husband Maciej for his constant love and belief that nothing is impossible. Our discussions on low-pressure single-piece flow carburising have laid the foundations for the subject matter of this work.

I extend my thanks to three great practitioners of thermal and thermo-chemical treatment: Daniel Herring—for generously sharing his scientific and industrial knowledge and donating a large amount of literature of his own authorship, Prof. Richard Sisson—for working together as part of the CHTE team and providing an opportunity to observe what it is like to work in a scientific-industrial consortium in a US setting, and Dr. D. Scott McKenzie—for the opportunity to view global activities in thermal and thermo-chemical treatment from the perspective of ASM International. Daniel, Richard and Scott have been my benevolent guides in the world of global commercialisation of thermal and thermo-chemical treatment.

I would like to thank two women of science: Dr. Marzena Kraszewska and Prof. Anna Dobrzańska-Danikiewicz, the embodiments of uncompromising determination in the pursuit of scientific excellence, from whose attitude I was inspired.

I would like to thank my supervisors, Prof. Łukasz Kaczmarek and Prof. Jacek Sawicki, for creating favourable conditions for working on the book, and my friends, Dr. Aleksandra Jaszczak and Anna Gnatkowska, for their continued support.

During the work on Chap. 6, dedicated to computational techniques, I was accompanied by the memory of Dr. Jerzy Dobrodziej, Eng. I know that he would have taken up the challenge of writing this work before me if he had not passed away prematurely.

When writing the book, I used the original texts wherever possible. This was possible, in particular, owing to the Jagiellonian Digital Library, the Main Library of the AGH University of Science and Technology and the Podlasie and Lower Silesia Digital Libraries. I would like to thank the Staff of these institutions, whose work in digitising Polish old prints has enabled me to travel less during pandemic.

Symbols and Acronyms

β_C	Carbon transport coefficient
β_N	Nitrogen transport coefficient
γ'	Layer of γ' nitrides
ε	Layer of ε nitrides
ε	Strain
ε^e	Elastic strain
ε^p	Plastic strain
ε^{tr}	Transformation plasticity strain
ε^{th}	Thermal strain
ε^{df}	Diffusion induced strain of carbon/nitrogen
ϕ	Volume fraction of the transformed phase
ρ	Density
σ	Stress
σ^y	Yield stress
ω	Dissociation rate
c	Heat capacity
d	Distance
h	Heat transfer coefficient
k	Speed reaction factor
m	Mass
m_{N_2}	Nitrogen availability
p	Pressure
t	Duration, time
w	Weight fraction
x	Mole fraction
C	Carbon concentration
C_P	Carbon potential
CD_{Eff}	Effective case depth
CD_{Tot}	Total case depth
D	Diffusion coefficient
D_C	Carbon diffusion coefficient

D_N	Nitrogen diffusion coefficient
F	Flow
Fe_α	Iron allotrope (ferrite)
Fe_γ	Iron allotrope (martensite)
J	Flux
J_Q	Heat flux
K	Heat conduction coefficient
L	Latent heat due to phase transformation
M_f	Final temperature of martensite transformation
M_s	Initial temperature of martensite transformation
N	Nitrogen concentration
N_P	Nitrogen potential
R	Boltzmann constant
S	Surface
T	Temperature
V	Volume
σ	Standard deviation for population
r	Range
s	Standard deviation for sample
$\overline{x}$	Mean value for sample
C_p	Process parameter scatter coefficient
C_pk	Process parameter distribution position coefficient
V	Coefficient of variation
ADALINE	ADAptive LINear Element
AI	Artificial Intelligence
BEM	Boundary Element Method
CFD	Computational Fluid Dynamics simulation
CPS	Cyber-Physical System
CVD	Chemical Vapour Deposition method
DL	Deep Learning
DM	Data Mining methods
FDM	Finite Difference Method
FEM	Finite Element Method
FVM	Finite Volume Method
HPGQ	High-Pressure Gas Quenching
IoT	Internet of Things
KDD	Knowledge Discovery in Databases
k-NN	k-Nearest Neighbours method
LPC	Low-Pressure Carburizing
LPN	Low-Pressure Nitriding
LSM	Least Squares Method
ML	Machine Learning
MLP	Multi-Layer Perceptron neural network
NN	Neural Network
PVD	Physical Vapour Deposition method

RBF	Radial Basis Function neural network
SPC	Statistical Process Control
SVM	Support Vector Method

Chapter 1
Introduction

You cannot fight against the future.

Wiliam Gladstone, 1809–1898

Our civilisation is changing and developing, and with it the expectations of the societies of different countries are changing. The social transformations over the centuries are so significant that, in sociological and economic sciences, the successive stages in the development of societies have been given names of their own, in the likeness of industrial revolutions [1]. The term Society 1.0 is used to describe a prehistoric stage in which people lived in small communities, providing food for themselves by hunting or gathering edible plants. This was followed by the agrarian period (Society 2.0), in which people farmed the land, cultivating a grain culture and becoming an agricultural society. This period saw the emergence of statehood and money as a means of exchanging resources. The amount of land owned became a measure of wealth. The turn of the seventeenth and eighteenth centuries brought further social transformations, linked to the Industrial Revolution (Society 3.0), concentrating multitudes of workers in newly established urban agglomerations. The invention of the steam engine and the introduction of electrification ensured that production lines operated around the clock; man's daily routine therefore changed radically to suit shift work in factories. The computerisation and robotisation of production lines in the second half of the twentieth century has again transformed the way society lived, making it an information society (Society 4.0). Data processing has become as important as matter processing—information gathering and exchange have become values in themselves, providing unlimited access to knowledge for people all over the globe. Today, data collected in the cloud and processed by artificial intelligence algorithms shape our daily lives, for example, by providing automatic translation of texts on the Internet, navigation on a mobile phone, searching for information on Google, or even controlling the home washing machine, fridge and oven. Moving forward, in 2016, the first country, Japan, signalled the need for another transformation improving the social life, postulating that this is only possible through the effective use of large-scale data (Society 5.0).[1]

The Japanese now define the term Society 5.0 as a human-centred model of human relations (human-centric society). This model uses economic progress to solve social problems through a system and technology that strongly integrates cyberspace with physical (real) space. Thus, we are talking about a society characterised by a higher level of integration (interpenetration) of the two realities—the digital reality and the genuine reality—facilitating the embedding of cyberspace in the real world [2, 3]. Japan was the first to propose a new model of how society should operate in response to the problems of energy shortages and foreign imports of energy, limited natural resources and an unfavourable demographic structure. An important aspect of this solution is the use of artificial intelligence methods (AI).

The concept described above—of developing the functioning of society—is also impossible without the advanced development of materials and the development of techniques for their manufacture. A parallel philosophical shift in the approach to the problem of materials and their production is needed to satisfy the signalled societal needs, as indicated by a comparative analysis of materials development in the context of societal development (Table 1.1). As will be substantiated below, in the context of the social assumptions made, there is no alternative to the development of industrial product manufacturing other than in line with the concept of Industry 4.0. The state of knowledge presented in this study also supports this thesis.

While Japan is in the vanguard of some social crises, other countries also have cause for concern. The unfavourable shape of the demographic structure is also evident in Europe, as is trouble in the energy sector. However, Europeans seem to be open to novelty in the way they function as a society just as much as they declare themselves open to using new technologies. According to a study by the Digital Poland Foundation conducted in Poland, 26% of Poles are techno-enthusiasts, 54% are techno-sceptics, 6% are picky technology users and only 14% are technology rejecters. Every year there is a growing tendency amongst Poles to settle their tax obligations via the Internet, to use electronic banking or to use e-providence [6]. It seems, therefore, that the vision of Society 5.0 is already meeting with a favourable response.

Material 4.0 and Industry 4.0

The term Industry 4.0 is derived from D. Bell's concept of industrial revolutions (Table 1.2). The first industrial revolution was named for the process of technological, economic, social and cultural changes initiated in the eighteenth century in England and Scotland, which saw the transition of production from craftsmanship

[1] Society 5.0—a human-oriented society in which economic progress proposing solutions to social issues is balanced by a system offering a high integration of digital and real space.

Table 1.1 Development of societies in the context of the materials engineering development [3–5]

<table>
<tr><th colspan="2">State of social development</th><th colspan="2">State of materials engineering</th></tr>
<tr><td>Society 1.0</td><td>Hunters-gatherers-societies (from 70 000 years BC.); basic social unit: group; trade exchange—barter; no privacy; division of labour by gender</td><td rowspan="3">Selection by trial and error: chemical composition of materials, process conditions of manufacture, range of applications</td><td rowspan="3">Material 1.0</td></tr>
<tr><td>Society 2.0</td><td>Agricultural society (from 8 000 years BC.); statehood; basis of exchange: money; land as a basis of wealth; functional division of labour</td></tr>
<tr><td rowspan="2">Society 3.0</td><td rowspan="2">Industrial society (from XVII/XVIII w.); concentration of population in cities on a massive scale; rigid, complex division of labour</td></tr>
<tr><td rowspan="2">Idea and laboratory works: characterisation, laboratory prototyping, testing and validation, pilot production, life cycle assessment</td><td rowspan="2">Material 2.0</td></tr>
<tr><td rowspan="3">Society 4.0</td><td rowspan="3">Information society (from 1955, USA); most important work with and on information; middle class; networks of formal and informal links</td></tr>
<tr><td>Modelling of a hypothetical material with assumed functionality: characterisation, laboratory prototyping, testing and validation, pilot production, life cycle assessment</td><td>Material 3.0</td></tr>
<tr><td rowspan="2">Big data type of data on materials fabrication and properties; computational materials science using artificial intelligence methods (AI), machine learning algorithms (ML), multiscale modelling; virtual materials synthesis and characterisation, prototyping, testing and validation; life cycle assessment</td><td rowspan="2">Material 4.0</td></tr>
<tr><td>Society 5.0</td><td>Society of imagination (from the 10th of the twenty-first century); integration of real-world cyberspace on the basis of technology; global community's pursuit of prosperity; smart manufacturing; digital transformation in which informationalisation is dominant</td></tr>
</table>

Table 1.2 Industrial revolutions (according to D. Bell) [1, 4]

Name	Beginning	Main characteristic	Characteristic invention	Other inventions
Industry 1.0	1784	Mechanisation	Mechanical weaving loom (1784)	Steam engine, railway, drainage canals
Industry 2.0	1870	Electrification	Production line (1870)	Electricity, automobile, aeroplane, assembly line, telegraph, telephone, radio
Industry 3.0	1969	Digitisation	Programmable logic (1969)	Automation, computer, PLC, digital communication
Industry 4.0	1991	Networking/ Internet	Internet (1991)	Data, digitisation, data cloud, IoT, network, general communication

(manufactory) to mechanical factory (industrial) production. The steam engine developed in 1769 by J. Watt made industrial plants, hitherto closely tied to the courses of rivers providing water power to move machinery, locationally independent. The development of the so-called weaving shuttle in England in 1733 allowed to develop spinning machines for mass textile production (the introduction of the power loom improved fabric production by increasing productivity 40-fold). This was facilitated by the increasing mass production of steel: in 1730, the metallurgical industry developed pig iron smelting using coal, and around 1780—blast furnaces for smelting using coke [1].

The second wave of development (the so-called Second Industrial Revolution) occurred at the turn of the twentieth century, bringing sweeping advances in science, technology, mining and metallurgy. Electricity has become a common way of lighting homes and streets. The first electric tram appeared.[2] Electric lighting, the electric motor, the car and the aeroplane came into use. Mass production, catering for the demographic explosion of the urban population, was streamlined by production and assembly lines.

The third major industrial change began in the 1970s and was the first change associated with the development of high-tech industries (Industry 3.0). The symbol of the third industrial revolution became the industrial districts, whose location no longer depended on the distribution of raw materials or energy sources, but on a clean environment, proximity to universities and access to skilled labour. Microcontrollers controlling machines came into widespread use; progressive process automation and information systems for production planning and control (i.e. automation of

[2] The first tram line was launched in Halle (Germany) in 1891.

industrial production using information and communication technologies) became commonplace.[3]

The most prominent feature of the fourth technological revolution (Industry 4.0) is the increasing intensity of combinatorial innovation. The classical approach to technological change proposed in the mid-nineteenth century by the Austrian economist J. Schumpeter assumed that its first stage was an invention: a new product or process, with the invention being scientific and technical in nature and not extending beyond the walls of the inventor's laboratory or garage. Innovation is already an economic fact with an impact on the economy, even if it is niche in nature. Only diffusion, i.e. dissemination, according to J. Schumpeter, transforms innovation into a socio-economic phenomenon [8]. A more recent concept, developed by, inter alia, Stanford economist A. Brian, assumes that technology is not created in a linear fashion, but that "technologies consist of other technologies, they arise as combinations of other technologies" [9]. This mechanism also characterised previous technological revolutions, but at a much slower pace, in particular due to the speed of knowledge circulation. The spread of the Internet, the pinnacle invention of the third technological revolution, has accelerated and unprecedentedly expanded the circulation and exchange of knowledge between researchers and institutions from around the world. In addition, as H. Varian, Google's chief economist, argues, the acceleration is also due to the dematerialisation of the main 'substrate' of innovation activities, namely data [10]. The topic of the fourth industrial revolution is as fascinating as it is vast, and unfortunately goes beyond the scope of the issues covered in this book.

In a classic model of the Industry 4.0 concept, H. Kagermann et al. identify nine technologies—pillars that support the transition of industry from the contemporary model to the Industry 4.0, including smart manufacturing [11–14]. These are:

1. Internet of Things, IoT,[4]
2. Cloud Computing,
3. Autonomous and Robotics Systems,
4. *Big data* Analytics,
5. Augmented Reality,
6. Cyber Security,
7. Simulation,
8. System Integration,
9. Additive Manufacturing.

---

[3] This division into industrial revolutions is conventional and was proposed in the 1970s by the American sociologist D. Bell [1]. A different, but similar classification comes from the work of C. Perez *Technological revolution and financial capital* (2002). These are successively: the industrial revolution—the launch of the first factory using an improved loom (1771), the revolution of dissemination—the test of the locomotive called the Rocket and the spread of the steam engine (1829), the revolution of steel and the electrification of heavy industry—the opening of the Canegie Bessemeer steel factory (1875), the revolution of oil, the automobile and mass production—the launch of the Ford T production line (1908), the revolution of information and telecommunications—the advent of the Intel microprocessor (1971) [7].

[4] Internet of Things—a digital network formed by interconnected objects and at the same time a network that connects to the physical world.

It seems that the model of H. Kagermann et al. is incomplete: it identifies only one manufacturing technology, somehow deprecating all others, and practically focuses only on cyber-physical processes and systems (CPS) or intelligent system—a computer system in which the way it operates is controlled or monitored by computer algorithms, including artificial intelligence systems. It is an integration of computing, networking and physical processes. Embedded sensors, computers and networks monitor and control physical processes through feedback loops, where physical processes influence computing and vice versa. The economic and social potential of such systems is far greater than has been the case to date, and large investments are being made in technology development around the world, in which the other eight technologies can be included. There is no doubt that CPS systems are indispensable in the areas of computation and process monitoring, however, it seems that a better description of the reality around us is the conceptually broader Idea 4.0 described by L. Dobrzański et al. [4, 15] as a combination of four aspects:

1. material,
2. technological process,
3. control system (CPS),
4. technological device.

A material world capable of satisfying human needs cannot exist without the material and the process and equipment necessary for its technological production.

Smart factory

The development of high-tech industries has come to symbolise the revolution of the early twenty-first century and in it the concept of, in which cyber-physical production systems control physical processes, collaborate with each other, make decentralised decisions and communicate with each other and people in real time. Through cloud-based data processing, they enable access to internal (machine sensors), inter-operational (indirect automatic quality checks) and final data (diagnostic data, periodic reports), which can be exchanged between systems within the factory or made available to external audiences (user, national systems, authorities, etc.). Such an intensive exchange of data on many levels requires a network of intelligent communicating devices (Internet of Things), the sensing of virtually all devices and the implementation of software to ensure automatic decision-making and action by these devices (Fig. 1.1). The smart factory is becoming the backbone of the modern production process according to the principles of Industry 4.0, ensuring a high degree of automation, robotisation and computerisation of industrial production [17–19].

Such a massive undertaking generates huge amounts of data from a variety of sources, which from 1997 has come to be referred to as *Big data*.[5] In 1999, D. Laney,

[5] In 2020, it has been estimated that the amount of data in the world will reach 44 zettabytes, or 44 billion gigabytes, by the end of the year.

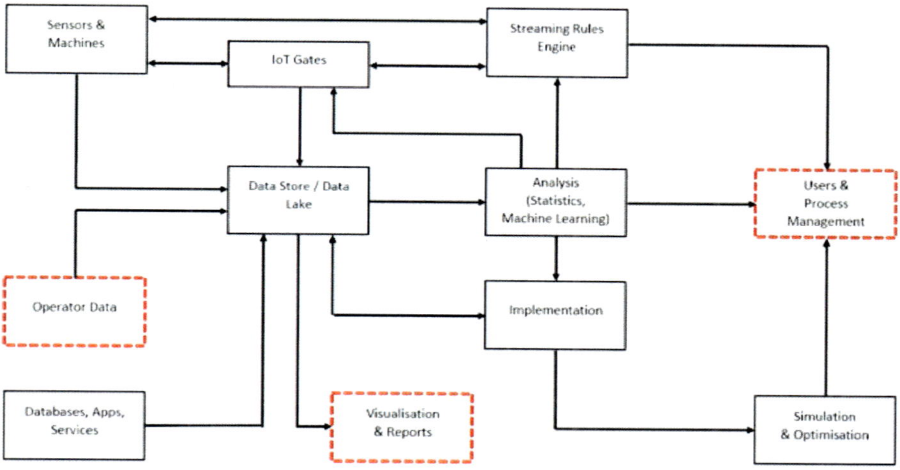

Fig. 1.1 Intelligent process control CPS system: dotted line indicates human contact points. Reprinted with permission from Ref. [16] Copyright (2021) StatSoft Poland

an analyst at the consulting firm Gartner, observing his clients' problems with data from different sources, its structure and varied formats, concluded that *Big data* is data of the *nV* type, which is characterised by: large volume, the velocity at which it is produced and variety. Over the next two decades, the list has lengthened to *10V*: in addition to those already mentioned, we discuss multidimensionality and inconsistency of *Big data* (variability), its relatively low reliability (veracity) as well as relevance and correctness (validity), its vulnerability to cyberattacks (vulnerability), its short usefulness in terms of the cost-effectiveness of archiving such large data sets (volatility), its visualisation challenges (visualisation) and its business value (value). Definitions of *Big data* equally often highlight that non-standard methods must be used to collect, process, analyse and visualise them. To quote a concise definition from Wikipedia: "*Big data* is the term used to describe the collection of data sets so large and complex that it becomes difficult to process them using database management tools or traditional processing applications" [10].

Apart from the obvious problems of processing and exploring such large and often unstructured data sets, the existence of an immeasurable body of knowledge brings with it the great potential of discovering new knowledge in data already collected. In the case of heat treatment, this refers to the modelling of phenomena and relationships knowledge of which is incomplete or even non-existent. The declining cost of conducting the computing (Fig. 1.2) allows to make new explorations in seemingly exploited data.[6]

[6] One FLOPS (FLOPS—*Floating Point Operations Per Second*) means one floating point operation performed per second. The processing power of today's processors is measured in gigaflops (GFLOP), a unit one billion times larger.

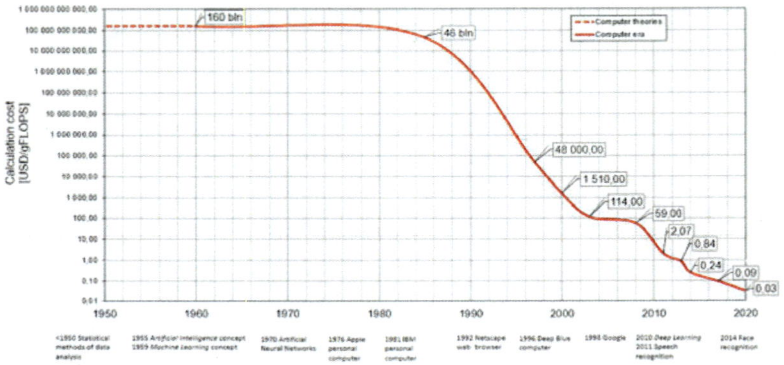

Fig. 1.2 Cost of running computer calculations in time perspective in USD/GFLOP. Based on data of Goldspot Discoveries Inc.

The basis of materials design is multi-criteria optimisation related to its classification into of one of the main engineering groups of materials or a group of natural materials, its chemical composition, manufacturing conditions, operating conditions and the way in which its waste is disposed of in the post-use phase, as well as price considerations related to the sourcing of materials, their processing into a product, the product itself and the cost of disposing of post-production and post-use waste, and the modelling of all processes and properties related to materials. The most advanced stage in the selection of materials so understood is referred to as the Material 4.0 concept (Table 1.1) [4]. In fact, engineering design encompasses three areas of activity: materials design, structural design of product forms and technological design of product manufacturing. A materials design methodology shaped in this way involves numerous activities related to the modelling and simulation of manufacturing processes and the prediction of operational properties of materials, the development of safe material technologies, the standardisation of tests of material properties and the development of a methodology for predicting the behaviour of new materials in use.

In heat treatment, the development and implementation of modelling and simulation technologies supporting heat treatment have been considered as one of the key directions for its development in the twenty-first century. This is particularly the case for thermo-chemical treatment, where it is virtually impossible to design and predict the effects of the process without computational support.

References

1. M. Waters, *Daniel Bell* (Routledge, London, 1995)
2. Japan-Government (ed.), *5th Science and Technology Basic Plan* (Japan Government, Tokyo, 2016)
3. J. Tworóg, P. Mieczkowski, A short story about Society 5.0, or how to live and function in the era of Economy 4.0 and 5G networks. Report for the Polish Chamber of Commerce for Electronics and Telecommunications, Digital Poland Foundation, Warsaw (2019). (in Polish)

4. L.A. Dobrzański, Role of materials design in maintenance engineering in the context of industry 4.0 idea. J. Achiev. Mater. Manufact. Eng. **96**(1), 12–49 (2019)
5. Y. Harari, *Sapiens. A Brief History of Humankind* (Harper, Glasgow, 2015)
6. P. Mieczkowski, M. Zubkowicz, P. Szkoła, Technology at the service of society. Will Poles become Society 5.0? Report Edition 2020, Digital Poland Foundation, Warsaw (2020). (in Polish)
7. C. Perez, *Technological Revolutions and Financial Capital: The Dynamics of Bubbles and Golden Ages* (Edward Elgar Pub, Cheltenham Glos, 2002)
8. J. Schumpeter, *The Theory of Economic Development* (Routledge, Oxfordshire, 1980)
9. A. Brian, *Nature of Technology* (Penguin, New York, 2010)
10. K. Śledziewska, R. Włoch, *Digital Economy. How New Technologies Change the World* (University of Warsaw, Warsaw, 2020). (in Polish)
11. H. Kagermann, W. Wahlster, J. Helbig, Recommendations for implementing the strategic initiative Industry 4.0, Final report of the Industry 4.0 working group. Report, Federal Ministry of Education and Research, Bonn (2013)
12. M. Hermann, T. Pentek, B. Otto, Design principles for Industry 4.0 scenarios: a literature review. Report, Technische Universität Dortmund, Dortmund (2015)
13. M. Rüßmann, P. Lorenz, P. Gerbert, M. Waldner, J. Justus, P. Engel, M. Harnisch, Industry 4.0: the future of productivity and growth in manufacturing industries. Technical report, Boston Consulting Group, Boston (2015)
14. W. Leong, J. Chuah, B. Tee (eds.), *The Mine Pillars of Technologies for Industry 4.0* (Institution of Engineering and Technology, Edison, 2020)
15. L.A. Dobrzański, A. Dobrzańska-Danikiewicz, L.B. Dobrzański, Effect of biomedical materials in the implementation of a long healthy life policy. Processes **9**, 865 (2021)
16. T. Demski, *Data Analysis and Artificial Intelligence for Industry 4.0 (Tutorial)* (StatSoft Poland, Cracow, 2021) (in Poland)
17. E. Hozdic, Smart factory for Industry 4.0: a review. Int. J. Modern Manufact. Technol. **7**, 28–35 (2015)
18. S. Wang, J. Wan, D. Zhang, D. Li, C. Zhang, Towards smart factory for Industry 4.0: a self-organized multi-agent system with Big Data based feedback and coordination. Comput. Netw. **101**, 158–168 (2016)
19. L.A. Dobrzański, A. Dobrzańska-Danikiewicz, L.B. Dobrzański, J. Dobrzańska (eds.), *Synthesis and Characterization of Biomedical Materials* (MDPI, Basel, 2021)

Chapter 2
Modern Thermo-chemical Treatment

Someone is sitting in the shade today because someone planted a tree a long time ago.

Warren Buffett, 1930

2.1 Early Scientific Papers

The importance of treatments that can enhance the functional properties of steel was already understood in ancient Greece. The lack of knowledge about the phenomena of hardening iron alloys was replaced by the craftsmen of the time with experience passed down from generation to generation and intuitive experimentation with hardening conditions. Different cooling effects of oil and water were distinguished and the quenching temperature was determined by visual inspection of the surface colour. The first unconscious attempts at carburising and nitriding can also be seen in the activities of Antiquity [1–4]. Biological fluids were readily used as a cooling agent in the Middle Ages; some, such as blood and urine, were beneficial in slowing the flow of heat during rapid cooling. Other organic substances, such as mustard, vinegar, resin, honey or leaf juice, probably caused transient deposits on the surface of the heated steel, which slowed cooling and reduced the likelihood of quench cracks. As highlighted by the historians of the subject, this was a period when a number of magical procedures were engaged into the practice of quenching, most likely to protect the artisanal profession of blacksmithing, customarily passed from father to son, from competition. This would explain recipes requiring the use of exotic cooling agents such as may dew, worm or snail juice, human blood and dragon blood [5–7].

Heat treatment as an object of scientific interest emerged in the sixteenth century. In 1639, R. Descartes argued that red-hot steel is filled with particles of fire which, when quenched in water, are replaced by much smaller particles of water. He believed that particles of water were too small to completely fill the pores of the metal, so the pores narrowed and the steel itself became harder [5, 6, 8].

In 1671, J. Rohault presented the *freezing model*, based on Descartes' concept of particles. According to this theory, during heating, the metal particles were set in motion while the distance between the two closest particles decreased. The sudden cooling prevented the particles from returning to their original position, causing them to *freeze* and form a denser, strengthened metal. The final result of the whole thought process did not differ from R. Descartes' *model of contraction*, and only the narrowing of *pores* was attributed to a different cause [9, 10].

In 1680, Ch. Perrault, an architect of the Louvre, observed an increase in the volume of steel after hardening, which challenged the theory of shrinkage. In 1722 this led to coining a new concept by R. de Réaumur who believed that iron alloys consisted of two types of Cartesian particles: iron particles and a certain amount of *sulphur and salt* particles, which under certain conditions diffuse into the former. The fire was supposed to drive the smaller *sulphur and salts* out of the iron particles and the cooling prevented them from re-entering. The *sulphurs and salts* were supposed to *freeze* between the iron particles, which would harden the metal and cause it to increase in volume. However, R. de Réaumur never defined the nature of the particles of *sulphur and salt*, nor was he able to explain differences in the hardness of alloys as a function of their particle content [3, 5, 6, 10]. In parallel (in 1774), the Swedish chemist S. Rinman outlined the role of an unspecified substance in iron alloys, which he called *plumbago*. This substance, which we now know was carbon, was studied further by T. Bergman, a compatriot of S. Rinman, who determined its content in iron, wrought iron and steel. He attempted to explain the importance of *plumbago* in quenching with R. de Réaumur's freezing model, but he too was unable to explain the differences in hardness after quenching between steel and iron in terms of the different contents of this substance in the two metals. The substance under study was finally named carbon around 1800 [3, 6].

In Poland, the scientific literature pertaining to the manufacture and processing of iron alloys was initiated in the second half of the eighteenth century, during the reign of Stanislaw August. In 1781, K. Kluk published a two-volume work *Searching, knowing and using more useful fossils*, from which comes the first Polish description of cementation, or carburising [11, 12]. A year later, J. Osiński presented the work *Description of Polish iron factories* (1782), in which he included a number of historical information about Polish ironmaking and presented a statistical study of the iron production of the time and its economic aspects [13, 14].

As underlined by J. Vanpaemel, the seventeenth- and eighteenth-century models were purely academic and not applicable in practice. None of these models were able to explain the relationship between the final product of quenching and the quenching temperature and quenching factors. These models were primarily qualitative in nature, so it was not possible to verify or use them quantitatively [10]. Their shortcomings were mainly due to the inability to detect the basic allotropic forms of iron by chemical analysis alone, which was the current research technique at the time. It became possible to observe the structural effects of metals and measure some of their physical properties in the 1860s and this almost immediately led to further advances in fundamental research.

2.2 Industrial Revolution of the Nineteenth Century

The Industrial Revolution, which began with the English textile industry, caused the growth in importance of mining, steelmaking and metallurgy. Its full flowering (the so-called Second Industrial Revolution) occurred in the second half of the 19th and early twentieth centuries and was driven by the rapid development of science, accompanied by the emergence of new technical developments. The history of heat treatment development, documented in this period, and of the associated theory of diffusion, is a combination of the life stories of dozens of researchers, along with their wealth of experiences and life choices making up the human fates. An extensive review of the biographies of the researchers involved in, in particular, heat treatment, was presented by H. Mehrer in his book *Diffusion in solids* (2007) and by J. Vanpaemel in his publication *History of the hardening of steel* (1982), and I have taken some of the mentioned facts in particular from the works of these researchers [10, 15].[1]

In 1868, a professor at the Artillery Academy in St. Petersburg, D. Chernov, elaborated the issue of phase transformations of steel (he introduced the concept of critical temperatures) and determined them as a function of carbon content. The effect of carbon on the position of the critical points was presented graphically by making a first rough sketch of the *iron-carbon diagram* [21, 22].[2] The following year, G. Gore observed a momentary elongation for iron alloys at a critical temperature on cooling, while W. Barrett found in 1873 that this was accompanied by a sudden increase in temperature. These changes were attributed to allotropy, which, with the state of knowledge at the time, was understood as a molecular change regardless of the arrangement of atoms in the crystal [24]. The state of knowledge at the time in 1885 was summarised in the work of F. Osmond, an employee of the Le Creuzot armament factory in France, defining the theoretical basis of heat treatment, including the hypothesis of the allotropy of iron and its relationship to carbon, and the three allotropic varieties of iron: Fe_α, Fe_β and Fe_γ [25, 26]. At the same

[1] I have found the other mentioned facts in the works of L. Barr *The origin of quantitative diffusion measurements in solids: A centenary view* (1997), J. Philibert *One and a half century of diffusion: Fick, Einstein, before and beyond* (2005), as well as in the works of C. Tuijn, S. Rothman and A. Seeger [16–20].

[2] The first of the points was defined as the temperature below which the steel does not harden (point A_1). The transformation temperatures of iron and its alloys, i.e. the critical points, are assumed to be denoted after F. Osmond's notation by the letter A (French *arret*, temperature stop). The symbols A_e denote the equilibrium temperature (French *equilibre*, equilibrium), while the symbols A_c and A_r denote the transformation temperature for heating (French *chauffage*, heating) and cooling (French *réfrigération*, cooling), respectively. This distinction was necessary because of the thermal hysteresis and, in addition, indicated the direction of the transformation. The magnetic transformation of iron Fe_α is designated as A_2, although it is not accompanied by a change in its structure [23].

time, J. Arnold proposed his notation of the diagram of phase transformations and this notation has been preserved to the present day [27]. The now commonly known *iron-carbon system* was developed by H. Roozeboom in 1900, which, together with the knowledge of critical points, explained the mechanism of quenching phenomena quite well [28]. Critical points had been determined by 1900 for many alloys using dilatometric measurements, resistance measurements, thermocouples, photometric methods, spectroscopy and calorimetry for this. A number of structures were also observed for alloys and were catalogued in terms of composition and heat treatment [6, 10].

There are also the first reports of studies of diffusion phenomena in the scientific literature of the nineteenth century. Probably the first experimental works (quantitative studies of diffusion between 1828 and 1833) were carried out by T. Graham, professor of chemistry at Strathclyde University in Glasgow, author of the theorem now known as T. Graham's law: *"Diffusion or spontaneous mixing of two gases takes place by the exchange in the position of infinitesimally small volumes of gases whose volumes are not of equal size, being, in the case of each gas, inversely proportional to the square root of the density of that gas"*. A key point of T. Graham's work on diffusion in gases was its connection to the kinetic theory of gases by J. Maxwell and R. Clausius, since T. Graham's law can be attributed to the balancing of kinetic energies between molecules of different molecular masses. Diffusion was thus linked to the thermal movement of atoms or molecules, and the idea of the mean free path emerged in science. In the following years Graham extended his research to include studies of the diffusion of salts in liquids and studies of the absorption of hydrogen in metals and showed that diffusion in liquids is at least several thousand times slower than in gases [29–31].

In 1855, T. Graham's work on the diffusion of salts in water inspired A. Fick, who developed a mathematical framework for describing the phenomenon of diffusion, using an analogy between the law of thermal conductivity and Fourier diffusion [32, 33]. His approach was phenomenological and used a description of the continuum.[3] A. Fick's important contribution to the development of diffusion theory was the definition of the diffusion coefficient and the determination of its value for the diffusion of salts in water.[4] Works on the mathematical solutions of A. Fick's equations were pursued by J. Stephan and F. Neumann, who were amongst the first to recognise the importance of boundary conditions for the solutions of the diffusion equation [37].

---

[3] Interestingly, A. Fick is better known in the field of medical science, where he earned his doctorate with his thesis *Visual errors due to astigmatism* (1851), published a monograph *Medical Physics* (1856) and a textbook *Anatomy of the sense organs* [34].

[4] The works of T. Graham and A. Fick on diffusion were continued by W. Roberts-Austen, but as noted by H. Mehrer, it is surprising that in their works the mentioned researchers did not recognise the temperature dependence of the diffusion coefficient [15, 35] The temperature dependence of reaction rates, diffusivity etc., now called *Arrhenius law* was probably formulated by the Dutch, Jacobus van't Hoff, the first winner of the Nobel Prize in Chemistry (1901), and now named after the Swedish scientist Svante Arrhenius, winner of the Nobel Prize in Chemistry in 1903 for his theory of electrolytic dissociation. S. Dushman and I. Langmuir postulated in 1922 that diffusivity in solids is also in accordance with this law [36].

At the end of the nineteenth century, the hardness of steel was explained by the existence of a hard compound of iron and carbon, *hardenite*, existing above the A_1 temperature, which underwent *freezing* during quenching. Moreover, F. Osmond believed that the cause of quenching was the transition of iron into the allotropic variety Fe_{β}, because the sudden cooling prevented the normal transition into iron Fe_{α}. Since experimental observations proved that hardenite predominated in high-carbon steels, while the allotropic effect (transformation of Fe_{β} into Fe_{α}) predominated in low-carbon steels, H. Howe introduced the term *martensite* in 1895, replacing the term *hardenite*.[5] It was thought that *martensite* was a high-temperature phase that was retained on sudden cooling, although some researchers challenged this theory due to a lack of explanation, so the same allotropic phase was non-magnetic at the quenching temperature and strongly magnetic after quenching [38]. H. Howe's colleague, A. Sauveur, suggested that the high-temperature phase is not a chemical compound, but rather a solid solution that decomposes during quenching into an aggregate of finely dispersed carbides embedded in a highly magnetic ferrite matrix. This structure was also referred to as martensite, but from then on the quenched structure was no longer considered to be a frozen high-temperature phase, but the result of its decomposition [10].

2.3 Thermo-chemical Treatment of the Twentieth Century

2.3.1 Development of Heat Treatment

In 1903, H. Howe introduced the term *austenite*, referring to a solid solution of carbon in iron and recognising that this allotropic form of iron that dissolves carbon is Fe_{γ} iron and not Fe_{β} iron.[6] Thus, the term *martensite* had already started to function only for the component in hardened steel, as a product of the transformation of the high-temperature phase [38].[7] Approximately 1910, it was thought that carbon acted as a brake, which, upon sudden cooling, retarded the normal decomposition of austenite into ferrite and cementite [39].

Until approximately the end of the 1920s, there were no reports in science that would not have been known in the previous century. According to M. Cohen, probably the main reason for this was a misunderstanding of the nature of allotropy, which was still interpreted as a molecular change [38]. The realisation that this is in fact a crystallographic change has become possible only in the era of diffraction studies (*X-ray diffraction*) and electron microscopy studies. By expanding the research

[5] H. Howe introduced the term martensite in honour of A. Martens, the pioneer of metallography.

[6] H. Howe chose the name austenite as proposed by F. Osmond, in memory of the English researcher W. Roberts-Austen. The name ferrite (Latin *ferrum*, iron) was also introduced by H. Howe [23].

[7] However, in the first two decades of the twentieth century, there was still much confusion about the terminology used, e.g. austenite of eutectoid composition was still called hardenite, i.e. a high-temperature compound.

workshop with new methods, it was quickly revealed that iron Fe_{γ} has the same structure as Fe_{α} iron and that the iron in the martensite is Fe_{α} iron. The decisive experimental evidence that martensite is a solid solution of carbon in iron Fe_{α} was the discovery in 1926–27 of a tetragonal spatially centred martensite lattice [40]. Diagrams of isothermal transformations and diagrams of continuous cooling were generated approximately 1930 for the transformation of austenite under imbalanced conditions. As E. McGannon believes, this was the first significant contribution of academic science to industrial practice. The developed diagrams formed the basis of new heat treatments (alloy quenching, isothermal annealing, isothermal quenching) and facilitated thermomechanical treatment [41].

Studies of the isothermal transformation of austenite in 1930–35 revealed the existence of three temperature ranges, with three types of structures and—as already presumed at the time—probably three different mechanisms for the formation of these structures.[8] Between 1932 and 1939, I. Mirkin developed the concept of high-temperature pearlitic transformation, in which the temperature dependence of the elementary processes of pearlitic transformation, i.e. nucleation and growth in the size of pearlite grains, was determined: the rate of nucleation and growth of pearlite grains. I. Mirkin's concept of the existence of fluctuations, i.e. continuous deviations from the mean condition, was also considered to be equally important.

In 1927, G. Kurdyumov observed the tetragonal structure of a martensite cell using X-ray structural analysis of [42, 43]. He also presented the effect of carbon content on the lattice constant and the degree of tetragonality of martensite and showed that martensite is a supersaturated solid solution of carbon in Fe_{α} iron [44]. The kinetics of martensitic transformations were studied in detail in the early 1930s. Experimental observations of the needle-like form of the precipitates with orientations in specific crystallographic planes, the occurrence and acceleration of transformations under mechanical stresses, the high rate of formation of martensite crystals and the lack of compositional changes during the transformation led researchers to conclude that the martensitic transformation differs from ordinary phase transformations [45]. S. Shteinberg formulated a hypothesis in 1930 that martensitic transformation is induced by internal transformations, and a few years later, it was demonstrated that it was impossible to apply thermodynamic theory to martensitic transformation. The study of the kinetics of martensitic transformation and some of its features (in particular, in 1937 M. Bigeev discovered the phenomenon of thermal stabilisation of austenite) was of great practical importance. A natural conclusion of the study was the new modernisation of heat treatment methods proposed by A. Gulyaev in 1937. In summary, the pre-war period was characterised by a wide range of research, mainly into the kinetic principles of transformation, which were important in the formulation of heat treatments. A theoretical basis for the analysis of metallurgical problems were the general laws of chemistry, which were needed to study the physical processes of crystallisation, melting, evaporation and condensation but did not take into account

[8] Of course, the phase transformations and the martensitic, bainitic and pearlitic structures are involved.

the influence of the actual factor of crystallography, upon which the modern theory of transformations relies [46].

A growing demand for materials with enhanced mechanical properties caused by the outbreak of the Second World War at the dawn of the 1940s highlighted the need to develop the theory of alloying; until then, largely empirical methods of alloy composition and processing had been used, which could not lead to the effective use of steel, particularly for building special machines. General principles for the influence of alloying on the behaviour of alloys during heat treatment and the way they are processed quickly began to emerge. Equally many papers have been devoted to the study of the influence of alloying elements on the course of phase transformations in alloy steels [47–49].

In 1947, G. Kurdyumov formulated the theory of martensitic transformation, in which he unequivocally demonstrated that martensitic transformation in steel has a thermodynamic origin and is a special type of phase transformation [43]. The isothermal transformation of austenite to martensite and the existence of *elastic* austenite crystals were predicted and experimentally confirmed [48, 50–55]. Additionally, the theory of tempering of hardened steel was developed in the 1940s and was based on studies of the effect of tempering temperature and time and the concentration of carbon in iron Fe_α. The nature of the transformations occurring during tempering was also investigated [43, 46]. In 1903, H. Howe introduced the term *austenite*, referring it to a solid solution of carbon in iron and at the same time recognising that this allotropic form of iron that dissolves carbon is Fe_γ iron and not Fe_β iron.[9] Thus, the term *martensite* had already started to function only for the component in hardened steel, as a product of the transformation of the high-temperature phase [38].[10] Around 1910 it was thought that carbon acted as a brake which, on sudden cooling, retarded the normal decomposition of austenite into ferrite and cementite [39].

Practically until the end of the 1920s, there were no reports in science that would not have been known in the previous century. According to M. Cohen, probably the main reason for this was a misunderstanding of the nature of allotropy, which was still interpreted as a molecular change [38]. The realisation that it is in fact a crystallographic change has only become possible in the era of diffraction studies (*X-ray diffraction*) and electron microscopy studies. It was quickly revealed, by expanding the research workshop with new methods, that iron Fe_γ has the same structure as Fe_α iron, and that the iron in the martensite is Fe_α iron. The decisive experimental evidence that martensite is a solid solution of carbon in iron Fe_α was the discovery in 1926–27 of a tetragonal spatially centred martensite lattice [40]. Diagrams of isothermal transformations and diagrams of continuous cooling were elaborated in 1930 for the transformation of austenite under imbalance conditions.

[9] H. Howe chose the name austenite as proposed by F. Osmond, in memory of the English researcher W. Roberts-Austen. The name ferrite (Latin *ferrum*, iron), was also introduced by H. Howe [23].

[10] However, in the first two decades of the twentieth century, there was still a lot of confusion about the terminology used, e.g. austenite of eutectoid composition was still called hardenite, i.e. a high-temperature compound.

As E. McGannon believes, this was the first significant contribution of academic science to industrial practice. The developed diagrams formed the basis of new heat treatments (alloy quenching, isothermal annealing, isothermal quenching) and facilitated thermo-mechanical treatment [41].

Studies of the isothermal transformation of austenite in 1930–35 revealed the existence of three temperature ranges, with three types of structure and—as already presumed at the time—probably three different mechanisms for the formation of these structures.[11] Between 1932 and 1939, I. Mirkin developed the concept of high-temperature pearlitic transformation, in which the temperature dependence of the elementary processes of pearlitic transformation, i.e. nucleation and growth in the size of pearlite grains, was determined: the rate of nucleation and growth of pearlite grains. I. Mirkin's concept of the existence of fluctuations, i.e. continuous deviations from the mean condition, was also considered to be equally important.

In 1927, G. Kurdyumov observed the tetragonal structure of a martensite cell using X-ray structural analysis of [42, 43]. He also presented the effect of carbon content on the lattice constant and the degree of tetragonality of martensite, and showed that martensite is a supersaturated solid solution of carbon in Fe_α iron [44]. The kinetics of martensitic transformations were studied in detail in the early 1930s. Experimental observations of the needle-like form of the precipitates with orientations in specific crystallographic planes, of the occurrence and acceleration of transformations under mechanical stresses, the high rate of formation of martensite crystals and the lack of compositional changes during the transformation led researchers to conclude that the martensitic transformation differs from ordinary phase transformations [45]. S. Shteinberg formulated a hypothesis in 1930 that martensitic transformation is induced by internal transformations, and a few years later demonstrated that it was impossible to apply thermodynamic theory to it. The study of the kinetics of martensitic transformation and some of its features (in particular, in 1937 M. Bigeev discovered the phenomenon of thermal stabilisation of austenite) was of great practical importance. A natural conclusion of the study was the new modernisation of heat treatment methods proposed by A. Gulyaev in 1937. In summary, the pre-war period was characterised by a wide range of research, mainly into the kinetic principles of transformation, which were important in the formulation of heat treatments. A theoretical basis for the analysis of metallurgical problems were the general laws of chemistry, which were needed to study the physical processes of crystallisation, melting, evaporation and condensation, but did not take into account the influence of the actual factor of crystallography, upon which the modern theory of transformations relies [46].

A growing demand for materials with enhanced mechanical properties, caused by the outbreak of the Second World War at the dawn of the 1940s, highlighted the need for developing the theory of alloying; up to then, largely empirical methods of alloys composition and processing had been used, which could not lead to the effective use of steel, particularly for building special machines. Concepts quickly

[11] Of course, the phase transformations and the martensitic, bainitic and pearlitic structures are involved.

began to emerge of general principles for the influence of alloying on the behaviour of alloys in heat treatment and the way they are processed. Equally many papers were devoted to the study of the influence of alloying elements on the course of phase transformations in alloy steels [47–49].

In 1947, G. Kurdyumov formulated the theory of martensitic transformation, in which he unequivocally demonstrated that martensitic transformation in steel has a thermodynamic origin and is a special type of phase transformation [43]. The isothermal transformation of austenite to martensite and the existence of *elastic* austenite crystals were predicted and experimentally confirmed [48, 50–55]. Also, the theory of tempering of hardened steel was developed in the 1940s: it was based on studies of the effect of tempering temperature and time and the concentration of carbon in iron Fe_α. The nature of the transformations occurring in tempering was also investigated [43, 46].

The development of the heat treatment theory of this period can be linked to the concepts of the structural characteristics of alloys prevailing at the time. It was known that the chemical composition of steel was not the only factor affecting the properties of steel, because it was observed that for the same chemical composition, the behaviour of the material varied from melt to melt. This difference was explained by chemical inhomogeneities in the main alloy component, grain size differences and chemical inhomogeneities resulting from heat treatment conditions. The concept of *structural compatibility* was introduced later on, taking into account the new concept of so-called physical inhomogeneity, postulating the existence of imperfections in the material structure. The concept of the nature of phase transformations in metals and alloys was closely linked to the development of the concept of the actual structure of alloys. M. Blanter reports that the beginning of this stage of the theory's development involved the experimental confirmation that each metal is characterised by a theoretical strength (dependent on the strength of the bonds between atoms in the ideal—i.e. defect-free—elementary cells that compose it). The existence of metals with theoretical strength supported the concept of the existence of defects in the crystal lattice, responsible for the low strength of real metals and alloys. The use of new research methods such as electron microscopy and "tracer" atoms has led to experimental confirmation of defects in the crystal structure of metals and alloys. The method of internal friction has made it possible to study the interaction of atoms of impurities and dislocations [46, 56–62].

The 1980s saw the first applications of computers in materials science and the availability of calculating machines increased with the development of microprocessors. However, the problem in metal science research at that time was not the lack of computers, but the lack of qualified research staff combining metal science knowledge with programming skills; ready-made software was not available, either. The obstacles were gradually overcome and the computer became an increasingly widely used tool in research into metals [63–65].

2.3.2 Development of Diffusion Theory

In 1827, a Scottish scientist and biologist, R. Brown, described the random movement of small particles in a liquid suspension. The plant he was examining under a microscope was *Clarkia pulchella*, a wild flower found on the Pacific Northwest coast of the United States. The pollen of this plant contained granules of about six micrometres in size, and it was their constant random movement in suspension that was the subject of R. Brown's observation [66]. No significant progress was made in understanding this phenomenon until the end of the nineteenth century, although developments in heat theory and kinetic theory stimulated new experiments and conjectures. As H. Mehrer notes in his book *Diffusion in solids* (2007), it was striking that the founders and developers of kinetic theory, J. Maxwell, L. Boltzmann and R. Clausius, never published anything on Brownian motion. Probably the reason for the lack of progress was that the main research of this period focused on velocities of particles, while the path of a small particle, at the scales of the length available from microscope observations, is an extremely erroneous curve. It was not until A. Einstein and M. Smoluchowski understood that velocity was not useful to measure in this context. A. Einstein published 4 papers on Brownian motion [67–70]. He was the first to understand that the fundamental quantity was not the velocity but the mean squared displacement of particles and related it to the diffusion coefficient.[12] His considerations were supplemented between 1906 and 1916 by M. Smoluchowski [72–74].

The beginning of the twentieth century shifted deliberations over the movement of small particles from liquids to solids. In 1926, J. Frenkel introduced the concept of *disorder* in the field of solid state physics and suggested that thermal excitation causes atoms to move from their regular places in the lattice to interstitial positions, leaving voids in the lattice (lattice disorders consisting of pairs of empty places, i.e. vacancies, and atoms in interstitial positions are today called a Frenkel defect) [75]. C. Wagner and W. Schottky generalised the notion of disorder for binary alloys, outlining the mechanism of defects formation, the locations of defects in the lattice and the relationships describing defectification and it is now widely accepted that point defects are an essential element mediating during diffusion in crystals [76].

A belief still prevailed in the 1940s that atomic diffusion in metals takes place through direct exchange or through ring mechanisms, which would suggest that in binary alloys the two components should have the same self-diffusion coefficient; meanwhile, already in the late 1930s and early 1940s, E. Kirkendall observed the unevenness of copper-zinc diffusion during the mutual diffusion between brass and copper [77–79]. The direction of mass flow was as would be expected if zinc diffused

[12] As the question of Brownian motion remained unsolved for most of the nineteenth century, the mathematician H. Pointcar described it in his book as one of the three fundamental unsolved problems in science. The two remaining problems were: *photoelectric effect* (explaining how ultraviolet light knocks electrons off the surface of a metal) and *luminiferous ether* (it was thought at the time that the Earth was moving in a medium that also carried light waves, but this could not be demonstrated). Inspired by H. Pointcar's book, A. Einstein developed and published an explanation of all three issues in April, May and June 1905 [71].

out of the brass faster than copper. This phenomenon has been confirmed in other binary alloys and the movement of inert tracers placed at the initial interface of the diffusion couple has been called the Kirkendall effect. This discovery was not appreciated until ten years after publication and is now considered the evidence that a diffusion mechanism exists in metals and alloys [15].

The 1930s brought an intensive development of research into self-diffusion; for a number of substances (gold, silver, copper, zinc, etc.), the dependence between the diffusion rate and temperature was confirmed, confirming Arrhenius' law [80–85]. The post-war period, on the other hand, allowed to use broadly radioactive isotopes produced in nuclear reactors and accelerators, studies of the isotope effect and studies of diffusion under hydrostatic pressure. Great improvements in the precision of diffusion measurements and in the accessible temperature ranges were achieved by using refined profiling techniques such as electron microprobe analysis, sputter sectioning, secondary ion mass spectroscopy, Rutherford back-scattering and nuclear reaction analysis. Methods were developed and widely applied, not directly based on Fick's law, to study atomic motion, such as the anelastic or magnetic after-effect, internal friction and impedance spectroscopy for ion-conducting materials. Completely new approaches making use of nuclear methods, such as nuclear magnetic relaxation (NMR), Mössbauer spectroscopy (MBS) and quasielastic neutron scattering (QENS) were successfully applied to diffusion problems [15].

The post-war development of information technology was also reflected in studies into diffusion. While diffusion at the surface of a solid body was recorded using scanning tunnelling microscopy, the movement of atoms inside the solid was still unavailable for direct observations. Since the diffusion of atoms within a material was a consequence of atomic or molecular jumps, a numerical approach to the mathematical problem of random walk allowed researchers to navigate between the diffusion coefficient and the jump distances and jump rates of the diffusing atoms. This has contributed to a significant growth in the use of computer modelling and simulation methods for diffusion processes in materials. Along with continuous modelling, which aims to describe complex diffusion problems using differential equations, atomic-level modelling, such as computations *ab-initio*,[13] molecular dynamics studies and Monte Carlo simulations, have played an important role as a means of gaining fundamental insights into diffusion processes [15]. Until 1950, the fact that diffusion occurred along the boundary of grains (grain-boundary diffusion) had been well documented, and the ratio of grain-boundary diffusion coefficients to diffusion coefficients in the crystallographic lattice in metals was estimated to be higher by several orders of magnitude [86]. In 1951, M. Fisher published a paper presenting the first theoretical model of grain-boundary diffusion [87], which started a whole area of quantitative studies over grain-boundary diffusion in solids [88]. Nowadays, diffusion across grain boundaries is widely recognised as a phenomenon of transport with fundamental and technical importance in normal polycrystals, and especially in nanomaterials [15].

[13] I.e. quantum chemistry computational methods.

A. Lidiard published the so-called *five-frequency model* in 1955, which established a theoretical framework for the diffusion of the dissolved substance and the solvent in diluted alloys and allowed to calculate correlation coefficients for the diffusion of the dissolved substance and the solvent. The model was also successfully applied to the diffusion of foreign atoms in ion crystals [89].

In the 1960s, a Norwegian, L. Onsager, presented the reciprocity theorem, which states that the matrix of phenomenological coefficients that bind the flows and generalised forces of the transport theory is symmetric. This theorem also includes cross phenomena, such as the effect of a concentration gradient of one type of atoms on the flow of another atom or the effect of a temperature gradient on the flow of atoms of different elements, both of which may be relevant to diffusion processes [15]. In 1968, L. Onsager was awarded the Nobel Prize in Chemistry and the worldwide recognition for the importance of his research inspired subsequent researchers into the phenomenon of diffusion and the mechanisms describing it at the end of the twentieth century [90–92].

In summary, the advancing knowledge of diffusion phenomena had a strong utilitarian potential and resulted in a significant development of heat treatment in industrial applications. In addition, in the twentieth century this was influenced by the general development of industry and the warfare waged around the world. The growing demand for modern weapons, ships, military vehicles, aircraft and spacecraft meant that, in parallel with the quantity demands of production, the quality requirements were rising. War conflicts accelerated the competition in the race for scientific and technological advances, as the importance of the mechanical properties of manufactured objects (at low and high temperatures), their chemical properties and the reliability of their operation was recognised and understood [4]. From the point of view of this work, this has led to an intensive development of thermo-chemical diffusion treatments, in particular carburising and nitriding.

2.3.3 Development of Carburising

The process of steel hardening by carburising has been a process known since antiquity, where it was used to make steel for swords and knives.[14] Since Greco-Roman times, people knew what to do so that manufactured swords do not bend or break after a strong blow, although for centuries there was no awareness of what happens during this process [93]. A prototype of powder carburising was used in sixteenth century, the first written recipes of which are preserved in the writings of the Italian scientist G. della Porta [94], and which were used unchanged for the next two hundred years. At the end of the eighteenth century, a Polish naturalist K. Kluk wrote in his work *Searching, knowing and using more useful fossils* [11, 12]:

[14] To be precise, surface hardening by carburising actually consists of three separate processes: the process of enriching the surface of a steel component with carbon (carburising), hardening the components and tempering to reduce brittleness.

This is well refined iron: from which are derived all the earthly particles, that the iron particles may mingle better with each other, and form a harder mass. The better the iron itself, the better the steel will be made from it. Everyone knows how manifold the needs of steel are; iron is seldom fit for use without it. The axes themselves must be made of steel. Crafts and trades cannot do without steel: they also know how to harden it so much that iron itself can be cut with it. Steel [on the other hand] is made from iron in many ways. ***By remelting.*** *The iron is remelted, and the more imperfect it is, the longer it lingers in the liquid on the fire: then it is hammered through.* ***By hardening.*** *Iron is placed in the melting furnace in layers, pouring it with the hooves of various animals: and once it has melted, it is quenched with cold water. This is repeated several times.* ***By cementation.*** *Grind the coals coarsely, and take the 1st part; a half of wood ash; mix the bones, horns, hides of cattle, mix the 1st part of ash in a closed vessel for burnt ash. Have an earthenware vessel made, cylindrical: narrow and long: three inches longer, as the steel bars should be. At the bottom of this vessel, pour the gunpowder just mentioned on your finger thickly and crush it. Place straight the iron bars that you want to turn into steel, but make sure that they touch neither the vessels nor each other: cover them completely with gunpowder, fill them from the top: cover them and seal them. Put in an even fire: after 10 hours prime steel you shall have.*

The texts of G. della Porta and K. Kluk are identical.[15] From the point of view of phenomena, both describe a surface hardening process, introducing carbon into the near-surface zones of iron alloys by means of diffusive saturation, without introducing changes in the core. The effect was achieved by heating the objects at the temperature at which homogeneous austenite exists, in an atmosphere with a reproducible (though numerically unknown at the time) carbon potential and a predetermined time. According to K. Kluk's text, the process was used to harden the surface for low-carbon and low-alloy steels.

Powder carburising

In the middle of the nineteenth century, when steelmaking by melting iron ore became an important industrial process,[16] carburising lost its importance as a steelmaking process, but began to gain new importance as a surface hardening technique to improve the properties of machine and tool components [93]. The importance of carbon transfer was understood in the nineteenth century and the first industrial carburisation process was developed. In this technique, parts were placed in a box filled with carburising powder, which consisted of charcoal, anthracite coal or coke. The box was sealed and placed in a furnace at a temperature of 870–930 °C. A limited number of oxygen molecules in the box reacted with carbon to form carbon monoxide, which, reacting with a second carbon monoxide molecule, underwent a

[15] *Take the hooves of a cow, put them in the furnace to dry so that they can be pulverised. Mix some of this powder with exactly the same amount of table salt, broken glass and chimney soot. Grind them and store it until use in a wooden vessel hung over the fire, (...). Then prepare an iron tin (...) put the files inside by pouring them thoroughly with powder (...). Then put the lid on, seal the crevices with clay mixed with chaff so that the smoke from the powder does not escape, and place it on the glowing coals (...). When you presume that the powder has already worn off and burned, remove the tin from the fire with iron tongs and immerse the files in very cold water (...)* (della Porta, 1589).

[16] Previously, it was not possible to obtain sufficiently high temperatures in furnaces.

disproportionation reaction to finally form carbon dioxide and carbon; the reaction proceeded on the steel surface according to the so-called Rideal mechanism.

Between the wars, the workpieces to be treated were placed in dedicated boxes filled with powdered (oak, beech, birch) charcoal, mixed with catalytic carbonates (barium, sodium, etc.) to accelerate carburising. The whole was heated at a carburising temperature of 900–950 °C. The oxygen in the box reacted with the charcoal, causing it to gasify to carbon monoxide, which then decomposed according to the Boudouard reaction to carbon dioxide and carbon diffusing deep into the steel [95]:

$$2C + O_2 \rightarrow 2CO \tag{2.1}$$

$$2CO \rightarrow CO_2 + C \tag{2.2}$$

Once the oxygen was consumed, carbon monoxide for the reaction was released from the charcoal gasification with barium carbonate and the resulting barium oxide reacted with the carbon dioxide produced from the decomposition of carbon monoxide, again producing barium carbonate:

$$BaCO_2 + C \rightarrow BaO + 2CO \tag{2.3}$$

$$BaO + CO_2 \rightarrow BaCO_3 \tag{2.4}$$

Powder carburising had the advantage of ease of implementation and low investment and operating costs. The disadvantage was the lack of control over the flux of carbon and over the carbon potential during the process, causing randomness in the structure of the layers obtained. Experiments with the initial carbon potential of the atmosphere led to creating a large number of carbides when the potential was too high, or to decarburisation of the surface when the potential was too low. In each case, the charcoal in the box was consumed during the process, causing an unintended reduction in carbon potential. The chemical composition of the atmosphere, after starting the process, could not be controlled or regulated [95]. In addition, as the process could not be mechanised and automated, the powder carburising technique, also known as box carburising, has lost its industrial relevance.

Salt carburising

The beginning of the twentieth century brought the development of carburising technology in molten chloride salts (also known as carburising in liquids or carburising in salt bath). The molten salt consists mainly of cyanides (NaCN, NH_4Cl, Na_2CO_3, NaCl); the activator caused the cyanide to dissociate, and the released carbon and nitrogen diffused to the surface of the part. The quantity ratio of cyanide to activator could be adjusted, affecting the carburising effect of the molten salts. Bath carburising gave greater control over the process than powder carburising, but the disadvantage of this method was its enormous environmental harm, which led to this method being abandoned with advances in technology [93, 96].

Gas carburising

J. Stanley of the Westinghouse Research Laboratories presented the equations for the carburising reaction in 1942 [97]:

$$Fe + 2CO \rightarrow C_{(Fe_\gamma)} + CO_2 \tag{2.5}$$

$$CH_4 + Fe \rightarrow C_{(Fe_\gamma)} + 2H_2 \tag{2.6}$$

where $C_{(Fe_\gamma)}$ denotes carbon introduced into austenite. He explained that carbon activity is the ratio of the partial pressures of carbon dioxide and carbon monoxide in the atmosphere:

$$k = \frac{p_{CO_2}}{(p_{CO})^2}, \tag{2.7}$$

as well as that the diffusion of carbon follows Fick's diffusion law:

$$C = C_S\left[1 - \varphi \cdot \left(\frac{d}{2\sqrt{D_C t}}\right)\right] \tag{2.8}$$

where C denotes the concentration of carbon at the point d, D_C is the diffusion constant, C_S is the carbon concentration at the surface, φ is the Gaussian error function, and t is the carburising duration [98].

In 1943, F. Harris, an employee of General Motors, noted the difference between the carbon concentration values calculated according to Fick's law of diffusion and those measured based on 24 practical operating data points [99]. He derived an empirical relationship for the thickness of the carburised layer on this basis:

$$CD_{Total} = \frac{31.6\sqrt{t}}{10^{6700/T}} \tag{2.9}$$

where CD_{Tot}—total thickness of carburised layer (inch), t—carburising duration (hrs), T—temperature (°F). The equation compensated for the difference between the measured data and the theoretically calculated data and had a functional use [98].

To sum up, the process of gas carburising in atmospheres from fuel gases was developed in the 1940s, initially using raw natural gas [100]. The atmosphere fed into the furnace chamber contained mainly methane, dissociating at the carburising temperature into carbon and hydrogen, and up to 10% carbon monoxide and dioxide, nitrogen and water vapour. The methane itself dissociated only partially to form a mixture of 25–40% of methane and 45–60% of hydrogen. A high-carbon potential was obtained from an atmosphere based on raw natural gas, but with low-carbon disposability. A characteristic feature of this method was that excessive surface carbon concentrations and large amounts of soot were obtained. This did not ensure correct results and created operational difficulties [95].

The importance of controlling the carburising atmospheres and carbon potential, and their impact on the final result of treatment, was quickly understood. The type, composition and flow rate of the carburising atmosphere dictated the value of the carbon potential obtained [101, 102]. R. Smith published a paper in 1946 on the equilibrium between carbon dissolved in iron and a carburising mixture (e.g. carbon monoxide-carbon dioxide, methane-hydrogen) [103, 104]. The concept of activity of carbon dissolved in austenite was already operating in science and knowledge of the influence of alloying elements on this activity was being developed. M. Blanter reported in 1948 that the diffusion coefficient of carbon in steel is independent of the concentration of carbon in that steel [105]. Fick's law was commonly applied. Deformations of the crystallographic lattice and diffusion directions in crystalline bodies were investigated. The state of the art of carburising in Poland in the 1950s was compiled by S. Kowal and W. Witek in their work *Steel carburising* (1957), who, while referring to the latest state of the science at the time, presented also its practical applications. They pointed to carburising in a solid medium widely used in industry (charcoal with addition of $BaCO_3$, $CaCO_3$, Na_2CO_3), in a gas medium (in gases containing carbon monoxide, methane or other hydrocarbons) and occasional applications of carburising in a liquid medium (in molten salts of Na_2CO_3 and NaCl type with the addition of SiC and CaC_2) [106].

The thermodynamic activity of carbon during carburising was determined in the 1970s as a function of interactions of alloying elements and its effect on controlling the carbon potential. The numerical values of parameters of the interaction function of silicon, aluminium, chromium, titanium, vanadium, molybdenum, tungsten, manganese and nickel were determined, characterising the influence of alloying elements on the thermodynamic activity of carbon for carburising in an endothermic atmosphere. The dependence of the carbon potential of the atmosphere as a function of the carbon activity coefficient in alloy austenite was determined [107, 108]. Mass exchange in the carburisation and decarburisation of steel carburised in endothermic gases was studied by V. Munts and A. Baskakov [109].

C. Dawes and D. Tranter presented an equation for the carburisation reaction and carbon potential in 1974:

$$H_2 + CO \rightarrow C_{(Fe_\gamma)} + H_2O \tag{2.10}$$

$$C_P = \frac{p_{H_2O}}{p_{H_2} \cdot p_{CO}} \tag{2.11}$$

where C_P denotes the carbon potential, $C_{(Fe_\gamma)}$ is the carbon incorporated into austenite, and p is the pressure of the gas in the atmosphere. They also demonstrated the importance of the reaction

$$CO + H_2O \rightarrow CO_2 + H_2 \tag{2.12}$$

occurring during carburisation and the ratio of the value of p_{CO_2}/p_{CO} and p_{H_2O}/p_{H_2} in controlling the carburising process. It was additionally demonstrated that the addition of propane as an enrichment gas reacts according to the reaction (2.13) to increase the carbon potential, while the reaction (2.14) directly affects the carburising [98, 110].

$$CO_2 + C_3H_8 \rightarrow 2CO + 2CH_4 \tag{2.13}$$

$$C_3H_8 \rightarrow C_{(Fe_\gamma)} + 2CH_4 \tag{2.14}$$

Unsuccessful attempts to modernise the process with the use of natural gas led to its abandonment and the introduction of processes with a controlled atmosphere.[17] An atmosphere of fuel gases (natural gas, propane) enriched with the same gas used to produce this atmosphere was introduced into a furnace chamber. At the process temperature, the enrichment gas dissociated and reacted with the components of the controlled endothermic atmosphere, producing a carburising atmosphere. In the 1950s and 1960s, exothermic atmospheres of the nitrogen and carbon monoxide-hydrogen-nitrogen types were used as protective atmospheres. In the 1980s, natural gas and technical propane were used to produce endothermic carburising atmospheres from fuel gases [95]. This was a major step towards technology with process control.

A variant of gas carburising was carburising in atmospheres of liquid organic compounds. Liquid organic compounds were introduced into the heating chamber of the furnace, where they were decomposed at the carburising temperature and the decomposition products constituted a carburising atmosphere. In the 1950s, the liquid used was paraffin or benzol, which was a mixture of benzene, toluene and xylene. Both substances were characterised by a variable chemical composition, making it difficult to obtain reproducible results. Numerous works by A. Moszczyński have revealed it was possible to regulate the carbon potential through the atmosphere from the decomposition of methanol with toluene [113]. Another method was methanol carburising using induction heating [114], but due to the highly toxic properties of methanol, both methods were used less frequently than the technical and economic rationale indicated.

Until the end of the twentieth century, gas carburising was considered the most industrially successful carburising method (in combination with integrated oil quenching), especially in mass production for the automotive industry. Thanks to controlled gas carburising, it was possible to regulate the carbon potential during the heat treatment cycle and create properly shaped carbon profiles in the surface layer of the workpiece. Since the introduction of controlled gas carburising, it has been possible to produce almost any shape of a carbon concentration profile [93, 115].

Carburising in a fluidised bed

Parallel to the development of gas carburising, the possibilities of carburising in a fluidised bed were investigated, where elements were immersed in a bed composed

[17] The gas was used again in the 1970s in reduced pressure carburising [111, 112].

of alumina or sand particles, lifted by a hot saturating gas flowing through the bed from below. Advantages of this solution were pointed out, such as reduced time of carburising in fluidised beds using propane-butane compared to gas, uniform saturation of external and internal surfaces, but disadvantages were also noted (strong soot-forming phenomena during saturation with gas, which blocked the process) [116–121].

Low-pressure carburising

A scientific discussion on the possible use of low-pressure carburising (LPC) (also known as vacuum carburising) arose in the 1960s [122]. Intensive research began in the 1970s, driven by the rising cost of natural gas and the hope of radically saving industrial gases by this method. The first work on low-pressure carburising in the pressure range from 100 to 600 hPa showed strong soot- and tar-forming phenomena and difficulties in obtaining a sufficient uniformity in carburising results [123–127]. Reducing the consumption of industrial gases to between 3 and 20 hPa using this method solved these problems and enabled to obtain a uniform thicknesses of the carburised layers in all elements of the charge and had a significant impact on the degree of homogeneity of composition and flow rate of the carburising atmosphere in the retort [128, 129].

In addition to temperature, pressure and time, the composition of the carbon-bearing atmosphere, its degree of homogenisation and the flow rate were of great importance for consistent results of the charge [130]. A. Moszczyński describes the feeding of natural gas at a pressure of approx. 1.3 Pa into the furnace chamber and then annealing the elements at a temperature of about 1,050 °C. The high temperature increased the solubility of carbon in austenite as well as the degree of methane dissociation; the high-carbon potential accelerated the carburising process without the risk of creating carbide precipitates, common in gas carburising at lower temperatures. A. Moszczyński pointed out that maintaining control in low-pressure processes is easier than in conventional processes and that in vacuum processes there is no internal oxidation due to the lack of oxygen in the carburising atmosphere. However, he indicated that when processes are carried out at temperatures above 1,000 °C, the growth of grains occurred, even in fine-grained steels [95].

Subsequent works performed over carburising with the use of propane revealed that its use causes difficulties in achieving homogeneous layers for densely packed charges and for components with complex geometries and in deep and blind holes. Problems with soot formation in the furnace chamber space also occurred. In 1999, W. Grafen and B. Edenhofer demonstrated that acetylene is a better chemically conditioned gas and does not cause the adverse phenomena characteristic of propane [131]. Low-pressure carburising for industrial use using acetylene was patented by K. Kubota in 1997 [132].

The low-pressure carburising technology has become increasingly important since the mid-1990s [133]. It has been and continues to be used industrially to enhance the properties of machine and tool parts in the automotive, aerospace, mechanical and other industries [132, 134, 135]. The main difference between low-pressure carburising and conventional gas carburising is that the low-pressure process uses gases derived from hydrocarbons as the source of carbon, instead of carbon monoxide. Consequently, the carburising atmosphere is anaerobic, allowing to completely avoid internal oxidation in the near-surface region of the carburised layer. The process has been modified and improved over the years, e.g. by replacing soot-forming carbon-bearing gases such as natural gas, propane or the propane-butane mixture with acetylene, ethylene and hydrogen. As the low-pressure carburising technology is mainly combined with high-pressure gas quenching instead of quenching in oil or in a water environment, it is seen as a very clean and environmentally friendly process [93, 115, 136].

Plasma carburising

Ion (plasma) carburising can be seen as low-pressure carburising in a partially ionised gas. The first hardness profiles generated by plasma carburising were presented by B. Edenhofer in 1972 [137]. The main process gas was methane, but the use of propane or mixtures of these gases with hydrogen or argon has also been reported [138–140]. A high-voltage electric field was used in the process itself, applied between the worked charge and the furnace wall, producing activated and ionised gas types responsible for transferring carbon to the workpieces [141]. The carburising process was followed by quenching in oil or quenching in gas under high pressure [142]. In 1991, B. Edenhofer reported on the implementation of the first six industrial installations operating in continuous production and executing plasma carburising, however, it did not ultimately find widespread industrial application [143].

Summary

Advances in the understanding of carburising processes have been closely intertwined with the development of diffusion theory in metals and alloys [54, 144–146]. S. Bokshtein reported in 1961 that the diffusion of carbon in Fe_α and Fe_γ proceeds non-uniformly. S. Bokshtein observed that carburising at temperatures of 550–950 °C enriches the grain boundaries in carbon, and that the diffusivity of carbon along the grain boundaries of ferrite is three to four times greater than inside the grains. He also noticed that alloying of steel alters the diffusion rate and distribution of carbon concentrations in a single grain and the carbon concentration gradient between the grain boundary and the grain interior is smaller for alloy iron than for unalloyed iron [147].

It was understood as early as in the 1940s that the correct structure of the carburised layer was not a sufficient condition to ensure the service life of machine parts in service, due to the deformations that can occur during hardening after carburising, and consequently more attention was paid in the 1950s to the stresses occurring in carburised layers [148]. It has been shown that the magnitude of residual stresses, their sign and distribution in the cross-section of carburised and quenched specimens depend on the depth of the hardened layer and the composition of the steel and that the residual austenite itself may undergo transformation during further processing or in the course of operation [149]. V. Sagaradze reported that the strength of carburised steel depends on the yield strength of the core and the residual compressive stresses, meanwhile, as the carbon concentration in the steel is growing, the yield strength of the core increases, while the compressive stresses decrease. He argued that, depending on the grade of alloy steel being carburised and the size of the component, there is an optimum carbon concentration to ensure the highest possible strength of the carburised steel [150]. It was also noted that gas quenching leads to significantly lower quenching strains compared to traditional quenching coolants (oil, water) [151].

The boundary conditions for carrying out carburising and the adverse surface and structural phenomena that occur when these boundaries are crossed were analysed at the turn of the 1960s and 1970s [152]. It was confirmed in gas carburising that high hardness and sufficient effective case depth are produced on steels by carburising to a temperature of approx. 1,000 °C, when fine martensite and evenly dispersed carbides are formed in the metallic matrix. An austenite zone was formed in the diffusion layer with a coarse network of carbides at the grain boundaries when carburising above 1,000 °C, leading to cracking when cooling from the carburising temperature (pre-cooling to quenching) or subsequent quenching [153, 154].

The first mathematical models of the gas carburising process giving up to 90% compatibility in structure prediction were appearing in the 1970s. B. Edenhofer reported in 1989 on an industrially working computerised system, *Carb-o-Prof*, which enabled to control the carbon profile constituted by the gas carburising process. With online calculations of carbon transfer/diffusion, the technique automatically regulated the two-step saturation/diffusion process, also taking into account unintended changes in carbon potential or temperature [155–158].

Today, carburising is one of the most widely used technologies to improve the performance properties of engineering surfaces [159].

2.3.4 Development of Nitriding

The creation of nitriding methods in the first half of the twentieth century was a response to the shortcomings of carburising methods at the time. The first practical development of nitriding in Europe in 1906 was made by A. Fry, a German metallurgist at Krupp Steel. While conducting a research project on the influence of alloying elements on the performance properties of steels, he noticed that the annealing of steels containing chromium, molybdenum, aluminium, vanadium and tungsten in ammonia resulted in a significant increase in surface hardness [160, 161]. As J. Zyśk notes, most researchers of the first quarter of the twentieth century pursued works with iron or carbon steel without achieving high hardnesses after nitriding. It is therefore believed that the rapid industrial success of nitriding was determined not so much by publishing the nitriding process, but by A. Fry's development of a composition of steel for nitriding containing chromium and aluminium. This enabled the immediate practical application of the theoretical results of his works, including the *iron-nitrogen diagram* he had developed [4, 160].[18,19]

Gas nitriding

A. Fry's process was introduced into industrial practice in the early years after the First World War and has been very successful in the aerospace, textile, railway, automotive and machinery industries [163]. Nitriding began to be used between 1926 and 1928 in Germany, France, America, England and the USSR. In Poland, it has also become an object of research works, inter alia, I. Feszczenko-Czopiwski and A. Wójcik introduced a new element of process control, i.e. the measurement of ammonia dissociation.[20] They also demonstrated that the degree of dissociation of ammonia depends on two factors: temperature and gas flow rate, with steel playing the role of catalyst [165, 166].

Compared to carburising, the newly developed nitriding was carried out at a temperature approx. 400 °C lower than the carburising temperature, and the steel after nitriding achieved a hardness of approx. 300–400 HV higher, with no deformations during processing. The hardening of the alloy occurred at process temperature and no hardening process was subsequently required. In addition, there was a perception in industrial practice that nitriding was much simpler in implementation than carburising. The advantages of nitriding were described in German literature as early as the 1920s; the high resistance of nitrided layers to wear, abrasion, corrosion and high surface compressive strength were particularly emphasised. The long process time (taking up to several tens of hours) was seen as a disadvantage [4, 163].

Floe nitriding

The 1930s and 1940s saw an intensification of theoretical and practical research into nitriding. A number of problem issues related to gas nitriding were clarified and new methods were developed [167–171]. However, a recurring effect of the formation of

[18] Interestingly, in the same year as A. Fry, the nitriding process was accidentally discovered by A. Machlet, an American Gas Company metallurgist in Elizabeth. He filed an application for a patent concerning heat treatment in 1907, in which he proposed replacing atmospheric air in the furnace with ammonia to avoid oxidation of steel parts. Shortly after he sent the patent application, he noticed that treating the parts in an ammonia atmosphere at elevated temperatures, *"skin, casing, shell or coating"* was formed around the part that was extremely difficult to corrode or tarnish [162]. The main differences between the methods elaborated by A. Machlet and A. Fry were that the A. Machlet process used hydrogen as a dilution gas to control the nitriding potential of both the gas and the steel, which in turn controlled the final surface properties. A. Fry controlled the process through alloying and focused on improving aspects such as core hardness and tensile strength. Although A. Machlet's development and patenting of the new nitriding procedure was important from a technological point of view, his work went largely unnoticed [163].

[19] Parallel to A. Fry *iron-nitrogen diagram* was developed by C. Sawyer, although he is rarely mentioned in the literature [164].

[20] In their works, I. Feszczenko-Czopiwski and A. Wójcik relied on J. Walsted's 1929 doctoral thesis.

a near-surface layer of iron nitrides on the surface of the nitrided steel was observed in these processes, which were identified in the studies as a mixture of two phases: γ' and ε.

In 1944, C. Floe, a professor at MIT in Boston, proposed a two-stage nitriding process that enabled to reduce the thickness of the near-surface layer of iron nitrides [167]. In the solution proposed, the first stage was carried out at a temperature of about 500 °C with an ammonia dissociation degree of 15–30% (i.e. in an atmosphere containing 70 to 85% of ammonia); under these conditions, a layer of iron nitrides was created on the surface of the nitrided steel, built up by the ε layer. In the second stage, the temperature was increased to about 560 °C and the degree of ammonia dissociation was increased to 75–85% (the atmosphere contained between 15 and 25% ammonia); this value was maintained until the process was complete. The growth of the thickness of the near-surface layer of iron nitrides formed in the first stage was inhibited at this stage. The two-stage process technique was dedicated to reducing the formation of the $\gamma' + \varepsilon$ zone and did not serve any other purpose [163]. Controlled gas nitriding, being a development of traditional gas nitriding, was widely used in the 1980s and 1990s. With the development of computer-based computing techniques, the possibility of computer control of this process has been explored through predictive calculations of the progress of diffusion and values of nitrogen potential [172, 173].

Salt nitriding

An alternative method to gas nitriding was salt nitriding developed in the 1920s, in which the nitrogen source was molten salts. The salt nitriding process used the phenomenon of cyanide decomposition to sodium and potassium cyanate followed by cyanide oxidation and cyanate dissociation. The nitrogen thus released was the nitrogen source of the process. This enabled to produce layers with a thin zone of nitrides that did not require grinding. The advantage of bath nitriding was the lack of brittleness of the layers produced, which allowed it to be applied to components made of carbon, alloy and high-speed steels working under high-pressure conditions of a point or linear nature. The disadvantage was the high toxicity of the process [174].

Ion nitriding

In 1932, in Germany, B. Berghaus patented and Wehnheldt developed the nitriding method, which was called *glow-discharge nitriding*. This was the beginning of works over ion nitriding. The nitriding agent in these processes was a nitrogen-hydrogen gas mixture, ionised in a vacuum furnace at a vacuum of 50 to 600 Pa by means of a strong glow discharge. Due to the high-energy action of the plasma, the working temperature during nitriding could be significantly reduced compared to salt bath and gas nitriding, allowing the process to be used for nitriding temperature-sensitive materials and components (high-porosity sinters, austenitic steels, etc.). These processes found application during the Second World War for the production of German military products (nitriding of cannon barrels), but were not widely used outside this country because they were seen as too complex, expensive and unreliable to guarantee consistent and reproducible results. The method of ion nitriding was not broadly accepted industrially in Europe until the 1970s and by the end of the twentieth century plasma nitriding technologies were in use all over the world [163].

Significant contributions to the development of ion nitriding were made by E. Roliński, E. Lunarska, T. Wierzchoń, J. Michalski, A. Marciniak and others [175–185].

Low-pressure nitriding

The need to reduce the supply of nitrogen during nitriding was the basis for developing new gas nitriding technologies and for the advancement of ion nitriding and low-pressure nitriding (also known as vacuum nitriding). These methods were more able to shape the microstructure of the surface layer as required [186–188].

The ability to control the amount of nitrogen delivered to the surface of the treated charge during the process was the essence of the reduced-pressure nitriding process in a dissociated ammonia atmosphere, known as *Nitrovac'79*, developed by Z. Haś. The advantage of this technology was that the low partial pressure of hydrogen had the effect of lowering the interaction of this element with the surface of the charge, and consequently the layers obtained by this method were more ductile compared to the layers obtained by the other processes [186, 187].

The concept of low-pressure nitriding since the 1960s saw the potential for better control of the nitrogen flux into the steel. In 1966, A. Smirnov and J. Kulesov presented a graph of the equilibrium *degree of dissociation of ammonia—pressure—solid phases of the iron-nitrogen system* and showed that lowering the pressure below atmospheric pressure causes a shift of the equilibrium towards the phases γ' and ε, containing more nitrogen [189]. V. Krylov et al. proposed the use of the steep part of the Langmuir isotherm (below 100 hPa) to monitor the nitrogen flux during low-pressure nitriding [135]. A change in the approach to control and monitor low-pressure nitriding from a thermodynamic mode, based on controlling the nitrogen potential, to a kinetic mode, based on controlling the degree of adsorption, was proposed this way. This fundamental statement was the basis for developing low-pressure nitriding procedures, which have found their way into practice [187]. It should be noted that all these processes were developed experimentally by trial and error, since at the turn of the twentieth and twenty-first centuries, theoretical models for nitriding under reduced pressure in a non-steady state for alloy steels still did not exist [190].[21]

Summary

A number of papers of a cognitive nature were published in the 1950s, 1960s and 1970s concerning the phase transformations occurring in nitriding and carburising [191, 192]. New types of short-term carbonitriding in cyanine-cyanide salts and gases

[21] The second reason for this situation was the lack of availability of sensors to analyse the outlet atmosphere at very low partial pressures of ammonia.

were developed, resulting in an increase in the resistance of machine parts to wear by friction and an increase in fatigue strength. Works were also conducted over sulphur nitriding in baths. The basics of nitriding were already well known and were commonly used to increase hardness, enhance abrasion resistance of component surfaces, and resistance to volumetric fatigue and surface fatigue (pitting), thus extending significantly the service life of machine parts being under high pressure. It was known that nitrides were the main hardening agent of steel, and since iron nitrides alone harden the material only to a small extent, an important role was attributed to alloying additives such as chromium, molybdenum, aluminium and vanadium, whose compounds with nitrogen enabled the material to harden to a level of 1,000–1,400 HV. It was also known that the properties of the nitrided layer depended on its structure, i.e. the thickness of the near-surface nitride layer and its chemical (phase) composition, as well as the thickness of the effective nitrided layer. However, in many cases, the presence of the zone of nitrides γ' and ε in the nitrided layer still reduced the service properties of the components through their high porosity and brittleness and their appearance in the nitrided layer most often necessitated removal by grinding. Ways were sought to reduce the adverse effects of conventional nitriding, inter alia by limiting the thickness of the nitride phases and reducing the proportion of the brittle ε phase by reducing the supply of nitrogen during the process. However, the reduction in nitride phases led to a reduction in hardness in the near-surface zone of the layer. To counteract this, ammonia began to be diluted with hydrogen or dissociated ammonia in the 1940s and 1950s, and with nitrogen in the 1960s [193]. Attention was drawn to the decarburising effect of hydrogen during the nitriding process, a phenomenon which causes an increase in the brittleness and a decrease in the hardness of the nitrided layers [194, 195].

In Poland, the first patent relating to gas nitriding appeared in 1977 and was developed by J. Tacikowski and J. Zyśk [196]. There were patent studies appearing in the following decades of the twentieth century mainly concerning atmospheric gas nitriding [187, 197, 198] and ion nitriding [199, 200], however, the truly numerous submissions were recorded in the early twenty-first century; these included both the methods of atmospheric gas nitriding [201–208], low-pressure gas nitriding [209], liquid phase nitriding [210–222] as well as ion nitriding [205, 218, 223–228].

As D. Pye aptly pointed out, by the end of the twentieth century, nitriding had reached maturity and had become a fully accepted, though sometimes not fully understood process [163].

2.3.5 Other Diffusion Treatments

H. Moissan observed in 1895 that it was possible to produce hard and abrasion-resistant surface layers by diffusion of boron into steel [229]. M. Czyżewski published a paper in 1915 on the cementation of iron by means of boron [230]. When studying Fe-B alloys, M. Czyżewski predicted that boriding of steel would have practical applications in technology. Between the two world wars, his research was taken up by I. Feszczenko-Czopiwski, focusing on the boriding of metal surfaces in hydrogen and vacuum atmospheres [231].

A significant advancement in boriding research occurred from the 1950s [232, 233]; Y. Lakhtin studied gas boriding of high-alloy stainless and heat-resistant steels and the influence of chromium and nickel on the hardness, depth and phase composition of the borided layer [234]. He demonstrated the feasibility of gas boriding of high-alloy steels, including austenitic steels. He observed that a two-phase layer is formed at the treatment temperature, consisting of a saturated solid solution (borides) and a solid boride layer. Y. Lakhtin confirmed their high hardness and wear resistance. I. Dukarevich et al. confirmed in the 1970s that there is a diffusion redistribution of carbon and alloying elements (chromium, manganese, silicon) in the electrolytic boriding process, which favours the formation of an intermediate layer. Carbon displacement results in decarburisation of the boride zone and saturation of the transition zone, while the presence of silicon promotes the formation of three zones in the borided layer (of borides, diffusion layer and transition zone), which differ considerably in composition and have a significant effect on the properties of the borided layer.

Originally, the term boriding was defined as the enrichment of the surface of a material with boron by thermo-chemical treatment. Using thermal energy and chemical reactions, boron atoms were adsorbed on the surface and inserted into the lattice of the base material and then diffused towards the core to form borides with the substrate atoms [235, 236]. However, considering the intensive development of boriding techniques, the term was extended to include other physical methods in which boron diffusion was less important. A general division of boriding techniques, including chemical and physical methods, was presented by K. Przybyłowicz in his work *Theory and practice of steel boriding* (2001) [237]. Today, boriding is used to enhance hardness, resistance to abrasive wear and corrosion in aggressive environments and to increase heat resistance of steel parts. This process involves the diffusion of boron atoms into the crystal lattice of the substrate material, resulting in the formation of a hard interstitial boron compound on the surface. The borided layer produced in the near-surface zone of the material can be in the form of a single-phase (Fe_2B) or a two-phase boride layer ($FeB + Fe_2B$). Boriding plays an important role in modern manufacturing technologies and is mainly implemented in two ways: by liquid electrolysis or gas boriding [238–241]. There are also numerous papers on the method of laser boriding [242]. Boriding enables to obtain a uniform hard layer throughout the depth from the surface to the end of the diffusion layer. The hardness achieved is many times higher than with any other surface hardening process. Wear, abrasion and surface fatigue properties are improved by combining high hardness and a low friction coefficient. Other benefits of boriding are that high hardness is maintained at elevated temperatures, corrosion resistance in an acidic environment and reduced consumption of lubricants. Borided layers are particularly resistant to abrasive wear in a loose abradant hence are used to produce tank tracks.

Apart from carburising, nitriding and boriding, which have found wide industrial applications, a number of other diffusion treatments have been developed in science and practice. The first attempts were made in the 1930s to cyanide tools in salt baths (USA, USSR) and in the late 1930s and early 1940s, low-temperature short-period nitrocarburising was introduced in an atmosphere made from ammonia with added hydrocarbons, which was used to produce thin anti-corrosion layers on carbon steel and cast iron products and to harden high-speed steel tools. After 1953, ferrite cyaniding (Tenifer, Durferrit) was introduced in Germany for the hardening of machine parts made of carbon steels, low-alloy steels and cast irons. After 1970, low-temperature cyaniding processes were gradually displaced by gas nitrocarburising, carried out in ammonia atmospheres with the addition of an endothermic gas atmosphere (*Nitemper*, Ipsen), exothermic gas atmosphere (*Nitoc*, Aichelin), endo-exothermic gas atmosphere (*Deganit*, Deguss) and exothermic gas atmosphere with methane, water steam and sulphur compounds. On the other hand, high-temperature carbonitriding processes have been used as an alternative to carburising [243].

In 1950, G. de Smet reported on a method to improve the wear resistance of surfaces, sulphonitriding [244, 245]. The method initially involved saturating the surface with sulphur, in a manner similar to the carburising process [246]. Further research has shown that only saturating the surface of steel parts with sulphur, carbon and nitrogen (sulphocyanation) produces the desired effects, as the combined effect of sulphur and nitrogen in the surface layer of the metal provides higher operating properties and wear resistance compared to saturation with sulphur only [247]. Sulphocyanation was used in what was then Czechoslovakia, France, England, the USA, Germany and Italy. It was widely used for surface refinement in piston rings, cylinder bushings for internal combustion engines, valve pushers and cast iron bushings for bearings. Sulphocyanine cast iron was replacing the previously used bronze (bushings, worm gears).

The anti-corrosion properties of chromium and silicon have been investigated since the 1940s, proving that the combined effect of chromium and silicon on the corrosion resistance of steel is stronger than the effect of each element separately [248]. This effect was confirmed even when only silicon was introduced and chromium was an alloying addition to the steel (chromium steel) [249]. The manufacture of corrosion-resistant chromium-silicon layers was patented and introduced in the 1950s [250, 251]. Both elements are also used separately, in chrome plating (saturating the surface layer of steel with chromium to increase corrosion resistance and heat resistance, and when chrome-plating high-carbon steels to increase hardness and wear resistance) and siliconising (saturating the surface layer of a part with silicon to increase corrosion resistance and acidity, used for low and medium carbon steels, as well as for ductile and malleable iron).

The diffusion saturation of surface layers with metals is called diffusion metallisation. The aim of its use is to produce hard layers, which are heat-resistant, resistant to erosion (atmospheric corrosion) and abrasive wear. Diffusion metallisation allowed this effect to be achieved cheaply and easily, as the desired effect was provided by layers as small as hundredths of a millimetre. Polish literature of the 1950s describes industrial applications of hot dip zincing, chromium plating, siliconising and alumination [106]. Many of these methods have evolved and are now classified outside of traditional galvanisation. Currently, metallic layers are produced by the methods of CVD (Chemical Vapour Deposition) or PVD (Physical Vapour Deposition)

[252, 253]. Titanising and vanadising applications for improving the performance properties of tool steels are also known [136].

2.4 Thermo-chemical Treatment at the Dawn of the Twenty-first Century

Heat and thermo-chemical treatment at the dawn of the third millennium was an important area of global industry, representing a key manufacturing step in almost every market sector: mechanical, tooling, automotive, rail and aerospace sector. Global heat treatment was dominated by surface hardening processes (i.e. carburising and post-carburising heat treatment). About 200,000 pieces of carburising equipment alone in the global industry were in operation during this period [174, 243]. Thermo-chemical treatment in the United States accounted for 65% of the market in 2000, being the basis of manufacturing processes for machine and equipment parts [243].

Today, heat treatment can even account for a major part of the total production costs of many components because it is capital intensive; the purchase of specialised treatment equipment requires significant capital investments, and the energy intensity of heat treatment processes makes them a considerable operating cost. For these reasons, heat treatment is becoming an important factor in determining the competitiveness and profitability of the manufacturing industry.

However, remaining competitive in a global market environment means a constant readiness to introduce innovations in the technological areas of the given sector. According to G. Totten (International Federation for Heat Treatment and Surface Engineering, IFHTSE), the global issue of heat treatment at the dawn of the twenty-first century has been defined by four aspects [254]:

1. Geometrical deformations after heat treatment, methods for their control and reduction and quality control of the product after heat treatment.
2. Optimum selection and use of cooling media.
3. Development and implementation of modelling and simulation technologies to support heat treatment, particularly thermo-chemical treatment.
4. Reducing the energy costs of carrying out heat and thermo-chemical treatment.

In 1999, to respond to the short- and long-term needs of the heat treatment industry, the American Society for Metals (ASM) created and implemented the concept of an industry and university research centre Center for Heat Treating Excellence (CHTE)—a consortium of the industrial and academic sector.[22] The initial successes of the chosen strategy has led Japan, Sweden, Australia and China to implement similar measures in their countries [255–257].

The works conducted by IFHTSE and ASM have resulted in a catalogue of specific courses of action recommended for rapid action, which is presented in Addendum Chap. 8 of this book. In 2007, the creation of databases of materials' properties (to

[22] The CHTE was located at Worcester Polytechnic Institute, USA.

increase modelling capabilities) and the improvement of vacuum carburising technology (low-pressure carburising), which was seen as having the greatest potential of all treatments, were considered as particularly important in terms of improving heat treatment technologies and materials. It has been shown that more intensive use of modelling to support the prediction of deformations and carbon concentration distributions in the carburised layer at the product design stage could help produce effective carburised layers with thicknesses reduced to a minimum, yet sufficient to ensure correct functioning of machine parts. Modelling of the heat treatment process that predicts quenching strains would allow the finishing treatment (stage of grinding to remove strains) to be reduced to a minimum, which would mean a reduction in raw material for excesses and a reduction in carburising duration and/or temperature. It was pointed out that computational support is of particular importance in non-equilibrium processes, where transient states of the process flow are present [258].

In the area of carburising improvement, great potential for reducing process durations was seen in high-temperature carburising processes, in which the Japanese industry already had some achievements. High-temperature carburising would allow for expediting the diffusion processes and for shortening the carburising cycle durations. It was believed that high-temperature rapid carburising was feasible for machining ultra-high-strength tool and die steels with a wide range of applications in manufacturing processes, provided solutions are developed to eliminate the grain growth problem that occurs at these temperatures [255].

Another area that was simultaneously identified as crucial and critical to the development of the heat treatment industry at the dawn of the twenty-first century was the development of the process modelling and numerical simulation [259]. Numerous analyses have ranked activities connected with the development of models as having the highest priority and cost-effectiveness for the heat treatment industry [260–262]. Issues related to this area are seen through many other problems, such as material selection, design of equipment structures [263, 264].

It is reported in regularly published review papers on the current state of thermal process modelling that models such as the microstructure model based on the chemical composition of alloy, atmosphere, temperature and treatment duration, or a model to predict residual stress and strain profiles would provide the heat treatment process engineer and designer with an invaluable tool to study important process parameters and their influence on the overall heat treatment process. In the 2000s, precise results of relatively simple shapes had already been obtained and successfully analysed for hardening, carburising, heat treatment and other processes [259]. However, the application of the models to more complex shapes, such as most manufacturing components, had to be preceded by the existence of an available database of high-temperature material properties. In addition, reliable heat transfer coefficients had to be generated and boundary conditions had to be defined [255]. In science, too, the lack of these tools has left gaps in the state of knowledge and prevented the solving of many scientific problems. For example, although it was known that full control of the nitrided layer was theoretically possible (e.g. producing only a solution layer or with a specific thickness of iron nitrides' layer), in practice this could only be achieved by numerical calculations supported by computer tools.

This study discusses the development of research works in the area of the three heat and thermo-chemical treatment problems identified in the literature at the dawn of the twenty-first century as the most significant problems in global heat treatment. Nevertheless, as this is a selection of only three issues out of many, it is necessarily a subjective choice made on the basis of the Author's professional experience and should be treated as such.

References

1. P. Jacquemart, J.F. Bois, *Notions de Technologie* (Delagrave, Paris, 1911)
2. P. Kosieradzki, *Heat Treatment of Metals* (PWT, Warsaw, 1955). (in Polish)
3. T. Wertime, *The Coming of the Age of Steel* (University of Chicago Press, Leiden, 1962)
4. J. Zyśk, *Development of Gas Nitriding of Iron Alloys* (Institute of Precision Mechanics, Warsaw, 2008). (in Polish)
5. C. Singer, *A History of Technology* (Oxford University Press, Oxford, 1954)
6. C. Smith, *Sources for the History of the Science of Steel: 1532–1786* (MIT, Cambridge, 1968)
7. J. Gordon, *The New Science of Strong Materials: Or Why You Don't Fall Through the Floor* (Pelican, London, 1974)
8. R. Descartes, *Oeuvres Philosophiques. Tome II - 1638–1642* (Editions Classiques Garnier, Paris, 1793). (in French)
9. M. Clagett, *Critical Problems in the History of Science* (The University of Winsconsin Press, London, 1959)
10. J. Vanpaemel, History of the hardening of steel: Science and technology. Le Journal de Physique Colloques **43**(C4), 847–854 (1982)
11. K. Kluk, *Seeking, Knowing, and Using Fossil Deposits*, vol. 1 (The Printing House of the King of the Republic of Poland, Warsaw, 1781). (in Polish)
12. K. Kluk, *Seeking, Knowing, and Using Fossil Deposits*, vol. 2 (The Printing House of the King of the Republic of Poland, Warsaw, 1782). (in Polish)
13. J. Osiński, *Description of Polish Iron Factories* (H. Małachowski, Warsaw, 1782). (Polish)
14. S. Płuszczewski, Literature of polish metallurgy until the mid-nineteenth century. Hutnik **8**, 330–338 (1936). (in Polish)
15. H. Mehrer, *Diffusion in Solids. Fundamentals, Methods, Materials, Diffusion-controlled Processes* (Springer, Berlin Heidelberg, 2007)
16. L. Barr, The origin of quantitative diffusion measurements in solids: a centenary view. Defect Diffus. Forum **143–147**, 3–10 (1997)
17. J. Philibert, One and a half century of diffusion: Fick, Einstein, before and beyond, in *Diffusion Fundamentals* (Universitatsverlag Leipzig, Leipzig, 2005), p. 8
18. C. Tuijn, On the history of models for solid-state diffusion. Defect Diffus. Forum **143–147**, 11–20 (1997)
19. S. Rothman, 100 volumes of diffusion - a personal view. Defect Diffus. Forum **99–100**, 1–24 (1993)
20. A. Seeger, A critical look at the mechanisms of self-diffusion in metals. Defect Diffus. Forum **143–147**, 21–36 (1997)
21. D. Chernov, Notes. *RTO Notes*, 7 (1868). (in Russian)
22. I. Gorbachev, Origin of the notations of critical points in steel. Metal Sci. Heat Treat. **10**(5), 341–343 (1968)
23. F. Staub, J. Adamczyk, Ł Cieślak, J. Gubała, A. Maciejny, *Metallography* (Śląsk, Katowice, 1978). (in Polish)

24. H. Pecheux, *Précis de Metallurgie* (B. Bailliere, Paris, 1868). (in French)
25. F. Osmond, J. Werth, Structure cellulaire de l'acier foundu. Comptes Rendus **100**, 450–452 (1885). (in French)
26. F. Osmond, J. Werth, Theorie cellulaire des proprietes de l'acier. Annales des Mines **8**, 5–84 (1885). (in French)
27. J. Arnold, The physical influence of elements on iron. J. Iron Steel Inst. **45**, 107–155 (1894)
28. H. Roozeboom, Fe-C phase diagram. The Metallographist **3**, 293–300 (1900)
29. T. Graham, Q. J. Sci. Lit. Art **27**, 74 (1829)
30. T. Graham, Philos. Mag. **2**, 175, 222, 351 (1833)
31. T. Graham, Philos. Trans. Roy. Soc. London **140**, 1 (1950)
32. A.E. Fick, Philos. Mag. **10**, 30 (1855)
33. A. Fick, Annalen der Physik und Chemie **94**, 59 (1855). (in German)
34. A. Fick, *Medical Physics* (F. Vieweg und Son, Brunswick, 1866). (in German)
35. W. Roberts-Austen, Phil. Trans. Roy. Soc. A **187**, 383 (1896)
36. S. Dushman, I. Langmuir, Phys. Rev. **20**, 113 (1922)
37. J. Stephan, Sitzungsberichte der Kaiserlichen Akademie der Wissenschaften **79**, 161 (1879). (in German)
38. M. Cohen, Strengthening of steel. Trans. Metall. Soc. AIME **224**, 638–657 (1962)
39. C. Macfarquhar, W. Smellie, *Encyclopedia Britannica* (Macfarquhar & Smellie, New York, 1910)
40. G. Kurdyumov, Mechanism of austenite-martensite transformation and behaviour of carbon atoms in martensite, in *Proceedings of the 1st International Symposium on New Aspects of Martensitic Transformations, May 10–12, 1976*, Kobe (1976)
41. H. McGannon, *The Making, Shaping and Treating of Steel* (United States Steel Corp, Pittsburgh, 1964)
42. N. Selyakow, G. Kurdyumov, N. Gudtsov, Zhurnal Prikladnoi Fiziki **2**, 51–54 (1927). (in Russian)
43. E. Vintaikin, G. Nosova, The founder of the theory of diffusionless (martensite) transformations on metallic alloys. Metal Sci. Heat Treat. **39**(1), 70–76 (1997)
44. G. Kurdyumov, E. Kaminskii, Zhurnal Prikladnoi Fiziki **2**, 47–48 (1929). (in Russian)
45. S. Shteinberg, S. Guterman, Metallurg **4–5**, 506–519 (1930)
46. M. Blanter, The development of heat-treatment theory. Metal Sci. Heat Treat. **12**(4), 328–332 (1970)
47. V. Sadovskii, M. Yakurvich, Reports of the academy of sciences of the USSR. Report 57, Academy of Sciences of the USSR, Moscow (1947). (in Russian)
48. G. Kurdyumov, Zhurnal Prikladnoi Fiziki **18**(6), 999–1026 (1948). (in Russian)
49. G. Kurdyumov, *Problems of Metal Science and Metal Physics*, vol. 3 (Metallurgiya, Moscow, 1949). (in Russian)
50. G. Kurdyumov, O. Maksimova, Reports of the academy of sciences of the USSR. Report 61, Academy of Sciences of the USSR, Moscow (1948). (in Russian)
51. G. Kurdyumov, L. Khandros, Zhurnal Prikladnoi Fiziki **7** (1949). (in Russian)
52. G. Kurdyumov, O. Maksimova, Reports of the academy of sciences of the USSR. Report 73, Academy of Sciences of the USSR, Moscow (1950). (in Russian)
53. G. Kurdyumov, O. Maksimova, *Problems of Metal Science and Metal Physics*, vol. 3 (Metallurgizdat, Moscow, 1952). (in Russian)
54. A. Gulyaev, *Heat Treatment of Steel* (Mashgiz, Moscow, 1953). (in Russian)
55. A. Gulyaev, V. Zelenova, *Fizika Metallov i Metallovienieje* **6**(5) (1958)
56. M. Krishtal, Y. Piguzov, S. Golovin, *internal Friction in Metals and Alloys* (Metallurgizdat, Moscow, 1964). (in Russian)
57. M. Grabski, R. Korski, Grain boundaries as sinks for dislocations. Philos. Mag. **22**, 707–715 (1970)
58. M. Krishtal, S. Golovin, *Internal Friction and Structure of Metals* (Metallurgiya, Moscow, 1976). (in Russian)

59. W. Łojkowski, M. Grabski, On material purity influence on the spreading temperature of grain boundary dislocations. Scripta Metallurgica **13**, 511–514 (1979)
60. M. Grabski, Mechanical properties of internal interfaces. Le Journal de Physique Colloques **46**, 567–579 (1985)
61. J. Wyrzykowski, M. Grabski, The Hall-Petch relation aluminium and its fependence on the grain boundary structure. Philos. Mag. A **53**, 505–520 (1986)
62. W. Łojkowski, J. Wyrzykowski, On the effect of dislocation spreading on grain boundary diffusion. Scripta Metallurgica **22**(1), 17–20 (1988)
63. M. Blanter, The computer in metal science. Metal Sci. Heat Treat. **28**(7), 465–471 (1986)
64. Y. Krupin, A. Kudrya, A. Melnichenko, Computer technologies in metal science. Metal Sci. Heat Treat. **41**(4), 168–173 (1999)
65. R. Gliner, Computer design of rational alloying of carburizing steels. Metal Sci. Heat Treat. **42**(7), 289–291 (2000)
66. R. Brown, Brief report of microscopic observations from June, July and August 1827 on particles contained in Clarkia Pulchella Pollen. Edin. New. Phil. J. **5**, 358–371 (1828)
67. A. Einstein, Die von der Molekularkinetischen Theorie der Warme Geforderte Bewegung von in Ruhenden Flussigkeiten Suspendierten Teilchen. Annalen der Physik **17**, 549 (1905)
68. A. Einstein, Zur Theorie der Brownschen Bewegung. Annalen der Physik **19**, 371 (1906)
69. A. Einstein, Zeitschrift fur Elektrochemie **13**, 98 (1907)
70. A. Einstein, Zeitschrift fur Elektrochemie **14**, 235 (1908)
71. J. Holt, *When Einstein Walked with Gödel: Excursions to the Edge of Thought* (Columbia University, Columbia, 2018)
72. M. Smoluchowski, Annalen der Physik **21**, 756 (1906). (in German)
73. M. Smoluchowski, Annalen der Physik **48**, 1103 (1915). (in German)
74. M. Smoluchowski, Physikalische Zeitschrift **17**, 557 (1916). (in German)
75. J. Frenkel, Z. Physik **35**, 652 (1926)
76. C. Wagner, W. Schottky, Z. Phys. Chem. B **11**, 163 (1930)
77. E. Kirkendall, L. Thomassen, C. Upthegrove, Trans. AIME **133**, 186 (1939)
78. E. Kirkendall, Trans. AIME **147**, 104 (1942)
79. A. Smigelskas, E. Kirkendall, Trans. AIME **171**, 130 (1947)
80. A. Sagrubskij, Phys. Z. Sowjetunion **12**, 118 (1937)
81. H. McKay, Trans. Faraday Soc. **34**, 845 (1938)
82. B. Rollin, Phys. Rev. **55**, 231 (1939)
83. W. Johnson, Trans. Americ. Inst. Min. Met. Engrs. **143**, 107 (1941)
84. P. Miller, R. Banks, Phys. Rev. **61**, 648 (1942)
85. C. Birchenall, R. Mehl, J. Appl. Phys. **19**, 217 (1948)
86. A. Le Claire, Philos. Mag. **42**, 468 (1951)
87. J. Fisher, J. Appl. Phys. **22**, 74 (1951)
88. R. Hoffman, D. Turnbull, J. Appl. Phys. **22**, 634 (1951)
89. A. Lidiard, Philos. Mag. **40**(1218) (1955)
90. J. Manning, *Diffusion Kinetics of Atoms in Crystals* (Norstrand Comp., 1968)
91. A. Allnatt, A. Lidiard, *Atomic Transport in Solids* (Cambridge University Press, Cambridge, 1991)
92. J. Philibert, *Atom Movements - Diffusion and Mass Transport in Solids* (Les Editions de Physique, Les Ulis, Cedex A, 1991). (in French)
93. B. Edenhofer, D. Joritz, M. Rink, K. Voges, Carburizing of steels, in *Thermochemical Surface Engineering of Steels. Improving Materials Performance*, ed. by E. Mittemeijer and M. Somers (Elsevier, 2015), pp. 485–553
94. G. Della Porta, *Natural Magic* (Neapol, 1589)
95. A. Moszczyński, *Gas Carburizing of Steel* (WNT, Warsaw, 1983). (in Polish)
96. E. Smol'nikov, A. Simonenko, Carburizing and decarburizing activity of salt baths with a graphite crucible. Metal Sci. Heat Treat. **16**(9), 727–731 (1974)
97. J. Stanley, Metal Prog. **11**, 849–855 (1942)

98. T. Watanabe, T. Hirata, *New Concept and Practical Operation of Carburizing and Nitriding* (Authors' Own Edition, Japan, 2015)
99. F. Harris, Case depth – an attempt at a practical definition. Metal Prog. **44**, 265–272 (1943)
100. W. Luty, H. Wilusz, Carburizing steel with natural gas 1. Przegląd Mechaniczny **3**, 12 (1958). (in Polish)
101. J. Chipman, S. Marshall, The equilibrium FeO+H2=Fe+H2O at temperatures up to the melting point of iron. J. Am. Chem. Soc. **62**, 299–305 (1940)
102. L. Darken, R. Gurry, The system Iron-Oxygen. The Wustite field and related equilibria. J. Am. Chem. Soc. **67**, 1398–1412 (1945)
103. R. Smith, Equilibrium of iron-carbon alloys with mixtures of CO-CO2 and CH4-H2. J. Am. Chem. Soc. **68**, 1163–1175 (1946)
104. R. Smith, Equilibrium of iron-carbon-silicon and of iron-carbon-manganese alloys with mixtures of methane and hydrogen at 1000C. J. Am. Chem. Soc. **70**, 2724–2729 (1948)
105. M. Blanter, Zavod. Lab. **3** (1948). (in Russian)
106. S. Kowal, W. Witek, *Steel Carburizing* (PWT, Warsaw, 1957). (in Polish)
107. B. Moiseev, Y. Brunzel, L. Shvartsman, Thermodynamic activity of carbon in recarburizing. Metal Sci. Heat Treat. **16**(1), 23–27 (1974)
108. S. Slavchev, V. Kluchitski, Thermoelectric method of controlling the carburizing process. Metal Sci. Heat Treat. **16**(2), 167–168 (1974)
109. V. Munts, A. Baskakov, Mass exchange in carburization and decarburization of steel. Metal Sci. Heat Treat. **25**(2), 98–102 (1983)
110. C. Dawes, D. Tranter, Heat Treat. Metals **11**, 121–130 (1974)
111. H. Westeren, The development of vacuum carburizing. Metallurgia Metal Form **10**, 390 (1972)
112. G. Prom, E. Jindel, Vacuum carburizing gives close carbon control. Metal Prog. **4**, 78 (1973)
113. A. Moszczyński, Systems for the regulation of generator-free carburizing atmospheres from liquid organic compounds. Biuletyn IMP **6**, 60 (1969). (in Polish)
114. A. Moszczyński, Induktive Aufkohlung von Armco-Eisen in flussigen Medien. Heat Treat. Metals **1**, 7 (1960). (in German)
115. P. Stratton, Carburising - looking back with a view to the future, in *Proceedings of 1st International Conference on Heat Treatment and Surface Engineering, May 16–18, 2013*, Chennai (2013), pp. 1–12
116. S. Iskhakov, V. Lomova, Yu. Strukov, Carburizing in a fluidized bed. Metal Sci. Heat Treat. **11**, 899–901 (1969)
117. V. Murav'ev, V. Kurbatov, N. Tyuteva, A. Govorov, Carburizing in a conducting fluidized bed. Metal Sci. Heat Treat. **13**(9), 773–775 (1971)
118. A. Baskakov, I. Kirnos, N. Zuikova, Surface processes in carburizing. Metal Sci. Heat Treat. **14**(1), 55–56 (1972)
119. T. Pumpyanskaya, V. Vinokurov, A. Baskakov, Rapid carburizing with natural gas in a fluidized bed. Metal Sci. Heat Treat. **14**(6), 493–497 (1972)
120. A. Baskakov, A. Zavarov, V. Svetlakov, Carburizing of Steel 10 in fluidized bed. Metal Sci. Heat Treat. **19**(1), 18–22 (1977)
121. A. Zavarov, S. Grachev, Heat and chemical heat treatment in a fluidized bed. Metal Sci. Heat Treat. **29**(10), 762–764 (1987)
122. C. Luiten, Wärmebehandlung mit Neuartigen Vakuumöfen in den Vereinigten Staaten. Heat Treat. Metals **18**(2), 172–175 (1963). (in German)
123. F. Eysell, Uber die Aufkohlung im Unterdruck Bereich, Verfahrensparameter und Anwendung. Elektrowaerme Int. **34**(6), 312–317 (1976). (in German)
124. F. Eysell, E. Heumüller, Entwicklungstendenzen auf dem Gebiet der Unterdruckaufkohlung. Heat Treat. Metals **34**(2), 83–89 (1979)
125. C. Luiten, F. Limque, F. Bless, Aufkohlen in Vakuum-öfen. Heat Treat. Metals **34**(6), 253–259 (1979). (in German)
126. F. Limque, F. Bless, Erfahrungen mit überdruck-Gasabschrecken Sowie Aufkohlen in Vakuumöfen. Heat Treat. Metals **35**(5), 238–244 (1980)

127. R. Chatterjee-Fischer, überblick über die Möglichkeiten zur Verkürzung der Aufkohlungsdauer. Heat Treat. Metals **40**(1), 7–11 (1985). (in German)
128. Y. Pourprix, J. Naudot, Carburizing under reduced pressure. Traitement Thermique **197**, 51–58 (1985). (in French)
129. H. Altena, Niederdruck-Aufkohlung mit Hochdruckgasabschreckung. Heat Treat. Metals **53**(2), 93–101 (1998)
130. F. Hoffmann, S. Dorn, P. Mayr, Aspekte des Unterdruck- und Plasmaaufkohlens. Heat Treat. Metals **49**(2), 31–39 (1994)
131. W. Gräfen, B. Edenhofer, Acetylene low-pressure carburizing – a novel and superior carburizing technology. HTM J. Heat Treat. Mater. **26**(4), 79–83 (1999)
132. T. Kubota, Vacuum carburizing method and device, and carburized products. Patent US 5,702,540, JH Corporation, Japan (1997)
133. B. Edenhofer, J. Bouwmann, Progress in design and use of vacuum furnaces with high pressure gas quenching systems. Ind. Heating **2**, 333–336 (1988)
134. E. Gronquist, V. Scotto, D. Taft, H. Westeren, Method of vacuum carburizing. Patent US 3,796,615, Hayes Inc., USA (1974)
135. V. Krylov, V. Yumatov, V. Kurbatov, Method for cementation of steel parts. Patent USSR 668978, Enterprise, Russia (1979)
136. P. Kula, *Surface Layer Engineering* (Lodz University of Technology, Lodz, 2000). (in Polish)
137. B. Edenhofer, Karbonitrieren in der Stromstarken Glimmentladung, in *Proceedings of 18. Härterei-Kolloquium*, Wiesbaden (1972)
138. B. Edenhofer, Fortschritte in der Prozessregelung beim Plasmaaufkohlen und Plasmanitrieren. Heat Treat. Metals **44**(6), 339–344 (1989)
139. Y. Yoneda, S. Takami, Ion carburising process for industrial applications, in *Proceedings of Vacuum Metallurgy* (Science Press, Princeton, 1977), pp. 135–156
140. P. Collignon, G. Hisler, H. Michel, M. Gantois, Study of carburising and carbonitriding by ion bombardment: comparision with conventional processes, in *Proceedings of Heat Treatment, May 6–7,1976*, London, 1976. The Metals Society, pp. 65–70
141. J. Dossett, G. Totten, *ASM Handbook*, vol. 4A. Steel heat treating fundamentals and processes (ASM International, Metals Park, 2013)
142. W. Gräfen, Plasma-carburising furnaces in European industry. Heat Treat. Metals **23**(4), 87–90 (1996)
143. B. Edenhofer, J. Conybear, G. Legge, Opportunities and limitations of plasma carburising. Heat Treat. Metals **18**(1), 6–12 (1991)
144. S. Gertsriken, I. Dekhtyar, *Diffusion in Metals and Alloys in the Solid Phase* (Fizmatgiz, Moscow, 1960). (in Russian)
145. M. Krishtal, Diffusion processes in iron alloys. Jerusalem **1963**, 90–133 (1970)
146. M. Krishtal, Diffusion in metals and alloys. Metal Sci. Heat Treat. **15**(7), 627–635 (1973)
147. S. Bokshtein, M. Gubareva, I. Kontorovich, L. Moroz, Peculiarities of the diffusion of carbon in iron. Metal Sci. Heat Treat. Metals **3**(1–2), 6–9 (1961)
148. U. Clanbite, Z.V.D.I. **91**, 577 (1944)
149. N. Vyaznikov, S. Ermakov, Residual stresses in carburized and quenched cases (on steel). Metal Sci. Heat Treat. Metals **1**(6), 41–44 (1959)
150. V. Sagaradze, Effects of carbon content on the strength of carburized steel. Metal Sci. Heat Treat. **12**(3), 198–200 (1970)
151. L. Karpov, Low-distortion quenching in inert gases. Metal Sci. Heat Treat. **15**(10), 857–862 (1973)
152. H. Grabke, Kinetics and mechanisms of the surface reactions between iron and gases during carburization and decarburization and nitriding and denitriding. Archiv für das Eisenhüttenwesen **46**(2), 75–81 (1975). (in German)
153. V. Belyakova, M. Alekseenko, Structure and properties of carburized martensite stainless steels. Metal Sci. Heat Treat. **11**(1), 32–34 (1969)
154. K. Lobl, G. Tuma, Carbide precipitation and carbide equilibrium in the Co-Cr-C system. Metal Sci. Heat Treat. **15**(1), 21–23 (1973)

155. I. Moskalenko, P. Ostrik, T. Kas'yanova, O. Krym, L. Ivanov, Gas carburizing of steels R6M5 and R6M5K5. Metal Sci. Heat Treat. **18**(11), 961–963 (1976)
156. L. Mikhailov, Calculation of the parameters of the process and the gas conditions during carburizing. Metal Sci. Heat Treat. **20**(7), 599–604 (1978)
157. B. Moiseev, Y. Brunzel', L. Shvartsman, Kinetics of carburizing in an endothermal atmosphere. Metal Sci. Heat Treat. **21**(6), 437–442 (1979)
158. B. Edenhofer, Adaptive carbon profile regulation in the gas carburising process. Heat Treat. Metals **16**(1), 22–27 (1989)
159. P. Stratton, S. Bruce, V. Cheetham, Low-pressure carburizing systems: a review of current technology. BHM Berg-und Hüttenmännische Monatshefte **151**(11), 451–456 (2006)
160. A. Fry, Stickstoff in Eisen, Stahl and Sonderstahl. Ein Neues Oberflachen Hartugsverfahren. Stahl und Eisen **4**(40), 1271–1279 (1923). (in German)
161. A. Fry, Process for hardening steel alloys. Patent DE 76108, Krupp Steel, Germany (1924)
162. A. Machlet, Hardening of treatment of steel, iron, etc. Patent US 1,092,925, Elisabeth, USA (1914)
163. D. Pye, *Practical Nitriding and Ferritic Nitrocarburizing* (ASM International, Materials Park, 2003)
164. C. Sawyer, Trans. AIME **69**, 798–822 (1923)
165. J. Walsted, *The case hardening of special steels with ammonia. A study of the nitriding process*. PhD thesis, Massachusetts Institute of Technology, Boston (1929)
166. I. Feszczenko-Czopiwski, A. Wójcik, Nitriding of steel by ammonia and aging phenomena. Research works of the state-owned armaments factory 1, Mining Academy in Cracow, Warsaw (1931). (in Polish)
167. C. Floe, A study of the nitriding process effect of ammonia dissociation on case depth and structure. Trans. AIME **32** (1944)
168. M. Bever, C. Floe, Case hardening of steel by nitriding, in *Surface Protection Against Wear and Corrosion* (ASM International, Metals Park, 1953), pp. 123–143
169. A. Minkevich, Thermochemical treatment of metals and alloys. Mashinostroenie **1**, 331 (1965). (in Russian)
170. Y. Sorokin, A. Minkevich, Nitriding steel in a mixture of nitrogen and ammonia. Metallovedenie i Termicheskaya Obrabotka Metallov **5**, 49–52 (1966). (in Russian)
171. A. Minkevich, Y. Sorokin, Investigation of steel nitrided in ammonia mixed with nitrogen. Metal Sci. Heat Treat. **9**(4), 309–311 (1967)
172. O. Bannykh, V. Zinchenko, B. Prusakov, V. Syropyatov, Advancement of nitriding in Russia. Metal Sci. Heat Treat. **41**(7), 287–291 (1999)
173. O. Bannykh, V. Zinchenko, B. Prusakov, V. Syropyatov, Development of nitriding in Russia. Forth period (1980 - present time): new directions in the development of LTCTT. Metal Sci. Heat Treat. **43**(3-4), 131–137 (2001)
174. P. Kula, Thermo-chemical treatment in a vacuum - prospects for development. Inżynieria Materiałowa **5**(112), 1999. (in Polish)
175. A. Marciniak, *Processes of heating and nitriding of the cathode during glow discharge diffusion treatment*. PhD thesis, Warsaw University of Technology, Warsaw (1983). (in Polish)
176. J. Michalski, T. Wierzchoń, Chemical vapour deposition of TiN layers in various glow discharge regions. J. Mater. Sci. Lett. **10**(9), 506–508 (1991)
177. J. Michalski, Ion nitriding of Armco iron in various glow discharge regions. Surface Coat. Technol. **59**, 321–324 (1993)
178. J. Mizera, R. Fillit, T. Wierzchoń, Residual stresses in nitrided layers produced on titanium alloys under glow discharge conditions. J. Mater. Sci. Lett. **17**(15), 1291–1292 (1998)
179. T. Wierzchoń, I. Ulbin-Pokorska, K. Sikorski, J. Trojanowski, Properties of multicomponent surface layers produced on steels by modified plasma nitriding processes. Vacuum **53**(3–4), 473–479 (1999)
180. T. Wierzchoń, I. Ulbin-Pokorska, K. Sikorski, Corrosion resistance of chromium nitride and oxynitride layers produced under glow discharge conditions. Surface Coat. Technol. **130**(2–3), 274–279 (2000)

181. E. Lunarska, K. Nikiforow, T. Wierzchoń, I. Ulbin-Pokorska, Effect of plasma nitriding on hydrogen behavior in electroplated chromium coating. Surface Coat. Technol. **145**(1), 139–145 (2001)
182. J. Sobiecki, T. Wierzchoń, J. Rudnicki, The influence of glow discharge nitriding, oxynitriding and carbonitriding on surface modification of Ti–1Al–1Mn titanium alloy. Vacuum **64**(1), 41–46 (2001)
183. E. Roliński, G. Sharp, The effect of sputtering on kinetics of compound zone formation in the plasma nitriding of 3% Cr-Mo-V steel. J. Mater. Eng. Perform. **10**(4), 444–448 (2001)
184. E. Roliński, Plasma-assisted nitriding and nitrocarburizing of steel and other ferrous alloys, in *Thermochemical Surface Engineering of Steels* (Elsevier, 2015), pp. 413–457
185. E. Roliński, Nitriding of titanium alloys, in *Heat Treating of Nonferrous Alloys*, vol. 4E, ed. by G. Totten, D.S. McKenzie (2016), pp. 604–620
186. Z. Haś, P. Kula, Nitrovac'79 – new technology of thermo-chemical treatment of machine elements and tools. Inżynieria Materiałowa **5**(16), 127–132 (1983). (in Polish)
187. Z. Haś, Method of subjecting components of machines, mechanisms and tools to gas nitriding. Patent PL 178378, Lodz University of Technology, Poland (1995)
188. N. Renevier, P. Collignon, H. Michel, T. Czerwiec, Low temperature nitriding of AISI 316L stainless steel and titanium in a low pressure arc discharge. Surface Coat. Technol. **111**, 128–133 (1999)
189. A. Smirnov, J. Kuleshov, Calculation of diluted ammonia nitriding reactions. Metallovedenie i Termicheskaya Obrabotka Metallov **5**, 45–49 (1966). (in Russian)
190. P. Kula, E. Wołowiec, R. Pietrasik, K. Dybowski, B. Januszewicz, Non-steady state approach to the vacuum nitriding for tools. Vacuum **88**, 1–7 (2013)
191. H. Grabke, Reaktionen von Ammoniak, Stickstoff und Wasserstoff an der Oberfläche von Eisen. Berichte der Bunsengesellschaft für physikalische Chemie **4**, 533–548 (1968). (in German)
192. H. Grabke, Evidence on the surface concentration of carbon on gamma iron from the kinetics of the carburization in CH4-H2. Metall. Trans. **1**(10), 2972–2975 (1970)
193. R. Chatterjee-Fisher, *Warmebehandlung von Eisenwerkstoffen, Nitrieren und Nitrocabrurieren* (Expert Verlag, Sindelfingen, 1986)
194. I. Feszczenko-Czopiwski, New trends in the cementation of iron with nitrogen. Przegląd Mechaniczny **1**(15–16), 561–567 (1935). (in Polish)
195. P. Kula, *Influence of hydrogen in the nitriding process on selected mechanical properties of 38HMJ steel*. PhD thesis, Lodz University of Technology, Lodz (1982). (in Polish)
196. J. Tacikowski, J. Zyśk, Method of gas nitriding. Patent PL 85924, Institute of Precision Mechanics, Poland (1977)
197. J. Trojanowski, J. Tacikowski, J. Zyśk, The method of preparing the surface of iron alloy machine parts and tools before their thermo-chemical treatment, especially gas nitriding. Patent PL 164982, Institute of Precision Mechanics, Poland (1994)
198. J. Panasiuk, J. Trojanowski, The method of nitriding structural steel elements. Patent PL 171363, Institute of Precision Mechanics, Poland (1997)
199. K. Miernik, Method of ion treatment, especially ion nitriding, of objects made of iron alloy. Patent PL 161571, Interdepartmental Scientific Center for Exploitation of Fixed Assets, Poland (1993)
200. K. Miernik, Method of ion treatment, especially ion nitriding. Patent PL 170008, Institute for Sustainable Technologies, Poland (1996)
201. M. Wysiecki, K. Szczeciński, J. Baranowska, Method of surface activation of iron and its alloys before diffusion nitriding. Patent PL 181175, Szczecin University of Technology, Poland (2001)
202. Z. Haś, Method of gas nitriding of machine parts, mechanisms and tools. Patent PL 191906, Lodz University of Technology, Poland (2006)
203. Z. Haś, B. Piotrowski. Method of nitriding machine parts and tools. Patent PL 192988, Lodz University of Technology, Poland (2006)

204. J. Panasiuk, A. Nakonieczny, J. Michalski, J. Tacikowski, I. Sułkowski, The method of gas nitriding of the inner surfaces of small and medium-caliber small arms barrels. Patent PL 203562, Institute of Precision Mechanics, Poland (2009)
205. J. Panasiuk, A. Nakonieczny, W. Dąbrowski, J. Trojanowski, Method of glow nitriding of inner surfaces of artillery gun barrels. Patent PL 206933, Institute of Precision Mechanics, Poland (2010)
206. Z. Haś, Method of nitriding machines and mechanisms. Patent PL 205169, Lodz University of Technology, Poland (2010)
207. V. Syropyatov, Device for catalytic gas nitriding of steel and alloys. Patent PL 211787, OOO Solnechnogoesky Zavod Termicheskogo Oborudovania Nakal, Poland (2012)
208. T. Żółciak, A method of gaseous nitriding and nitrocarburizing of ferritic alloy steels in fluidized bed furnaces. Patent PL 223176, Institute of Precision Mechanics, Poland (2016)
209. P. Kula, R. Pietrasik, E. Wołowiec, Method of nitriding tools made of iron alloys. Patent PL 219125, Lodz University of Technology, Poland (2014)
210. H. Hadj-Rabah, J. Terrat, A method of increasing the corrosion resistance of iron-containing metal elements by nitriding. Patent PL 177659, Andrezieux-Boutheon, Poland (1999)
211. U. Baudis, A method of producing corrosion-resistant surfaces of nitrided or nitrocarburized steel structural parts. Patent PL/EP 2235230, Durferrit GmbH, Poland (2017)
212. H. Chavanne, P. Maurin-Perrier, Bath of molten salts for nitriding steel products and method of its production. Patent PL/EP 2683845, Andrezieux-Boutheon, Poland (2019)
213. P. Magdinier, M. Desbouche-Janny, A method of surface treatment of a steel element by nitriding or nitrocarburizing, oxidation, and then impregnation. Patent PL/EP 3237648, Andrezieux-Boutheon, Poland (2020)
214. S. Srivastava, Cobalt-chromium-iron-nickel alloys susceptible to increasing strength by nitriding. Patent PL/EP 1900835, Haynes Int., Poland (2009)
215. T. Pełczyński, K. Lenik, Structural alloy steel for nitriding. Patent PL 212337, Lublin University of Technology, Poland (2012)
216. J. Szelc, Cz. Janczur, S. Róg, B. Stolarski, A. Citkowicz-Jurkiewicz, T. Jarmal, J. Minecki. A device for thermo-chemical treatment in a gas atmosphere of steel and cast iron elements, especially for treatment by nitriding. Patent PL 159142, Cracow University of technology & Elterma, Poland (1992)
217. J. Panasiuk, A. Nakonieczny, Vacuum, glow, horizontal furnace for thermo-chemical treatment of metals, especially ion nitriding. Patent PL 60848, Institute of Precision Mechanics, Poland (2004)
218. J. Panasiuk, A. Nakonieczny, I. Sułkowski, J. Tacikowski. Soak furnace for diffusion treatments, especially for plasma nitriding. Patent PL 61396, Institute of Precision Mechanics, Poland, (2005)
219. J. Trojanowski, J. Panasiuk, A. Nakonieczny, Z. Łataś, Horizontal glow furnace for thermo-chemical diffusion treatments, especially for ion nitriding. Patent PL 61573, Institute of Precision Mechanics, Poland (2005)
220. J. Trojanowski, J. Panasiuk, A. Nakonieczny, J. Tacikowski, Hood glow furnace for ion nitriding. Patent PL 61706, Institute of Precision Mechanics, Poland (2005)
221. H. Lankes, P. Haase, D. Miller, Method of manufacturing a retort for a nitriding furnace and a retort. Patent PL/EP 3183378, IVA Schmetz GmbH, Poland (2018)
222. I. Kocemba, W. Arabczyk, J. Rynkowski, A method of manufacturing a thermoelectric system for monitoring the processes of gas nitriding of iron powders and its alloys. Patent PL 236223, Lodz University of Technology, Poland (2020)
223. J. Panasiuk, A. Nakonieczny, J. Trojanowski. Glow nitriding method. Patent PL 193599, Institute of Precision Mechanics, Poland (2007)
224. M. Betiuk, J. Michalski, S. Kowalski, A. Nakonieczny, Method of glow nitriding of metal objects. Patent PL 217543, Institute of Precision Mechanics, Poland (2014)
225. M. Betiuk, J. Michalski, J. Tacikowski, Z. Łataś, T. Frączek, K. Maźniak, A method of glow nitriding or nitrocarburizing of the surface of metal objects. Patent PL 226062, Institute of Precision Mechanics, Poland (2017)

226. M. Betiuk, J. Michalski, J. Tacikowski, P. Wach, K. Burdyński, Z. Łataś, T. Frączek, A method of ion nitriding or nitrocarburizing of the surface of metal objects. Patent PL 230438, Institute of Precision Mechanics, Poland (2018)
227. S. Sobieski, A. Leciejewicz, B. Leciejewicz, J. Trojanowski, Method of ion nitriding of metal products and their alloys. Patent PL 236276, Institute of Precision Mechanics, Poland (2020)
228. M. Skóra, J. Trojanowski, Method of ion nitriding of tool surfaces made of alloy tool steels. Patent PL 236491, Solvera Gawel Technology S.A., Poland (2021)
229. H. Moissan, CR Hebdom Seances Acad. Sci. **120**, 74 (1895)
230. M. Czyżewski, *Zurnal Rus. Metał. O-twa* (1915), p. 645
231. I. Feszczenko-Czopiwski, *Boron Cementation (Boronization) of Iron, Nickel and Cobalt* (Mining Academy in Cracow, Cracow, 1925). (in Polish)
232. A. Minkevich, *Chimiko-tiermiczeskaja obrabotka stali* (Maszgiz, Moscow, 1950). (in Russian)
233. A. Minkevich, G. Ulybin, Chromizing and boronizing of steel with induction heating. Metal Sci. Heat Treat. **1**, 57–61 (1959)
234. Y. Lakhtin, M. Pchelkina, Boronizing of high-alloy steels. Metal Sci. Heat Treat. **3**(3–4), 111–114 (1961)
235. H. Kunst, O. Schaaber, Heat Treat. Metals **22**(1), 1–25 (1967)
236. H. Kunst, O. Schaaber, Heat Treat. Metals **22**(4), 275 (1967)
237. K. Przybyłowicz, *Theory and Practice of Steel Boriding* (Swietokrzyska University of Technology, Kielce, 2001). (in Polish)
238. A. Pertek, *Thermodynamic analysis of the BCl3-H2-Fe system and study of boride growth kinetics on selected iron alloys*. PhD thesis, Poznan University of Technology, Poznan (1980). (in Polish)
239. T. Wierzchoń, *Formation of an iron boride layer on steel under glow discharge conditions* (Warsaw University of Technology, Warsaw, 1986). (in Polish)
240. T. Burakowski, T. Wierzchoń, *Metal Surface Engineering* (WNT, Warsaw, 1995). (in Polish)
241. A. Pertek-Owsianna, *Shaping the structure and properties of iron boride layers obtained in the gas borizing process* (Poznan University of Technology, Poznan, 2001). (in Polish)
242. M. Kulka, *Current Trends in Boriding: Techniques* (Springer, Cham, 2019)
243. J. Michalski, Carburizing, nitrocarburizing, nitriding - development trends, in *Proceedings of 4th Seco/Warwick Seminar: Modern Trends in Heat Treatment'00*, Świebodzin (2000). (in Polish)
244. G. de Smet, Metallurgie et Construction Mecanique **82**, 12 (1950)
245. G. de Smet, Machine Moderne **45**, 503 (1951)
246. B. Jousset, Metal Prog. **65**, 76–80 (1954)
247. A. Kriulin, Sulfurizing and properties of sulfurized steel and iron. Metal Sci. Heat Treat. **2**(7), 389–395 (1960)
248. I. Kontorovich, *Heat Treatment of Steel* (Metallurgizdat, Moscow, 1945). (in Russian)
249. Z. Ordina, Diffusion of silicon in chromium steel. Metal Sci. Heat Treat. **2**(6), 349–350 (1960)
250. G. Kremer, Verfahren zur Herstellung von Korrosionswiderstandsfaehigen, Insbesondere Hitzebestaendigen Berzuegen auf Gegenstaenden aus Eisen und Stahl. Patent DE 892846C, Germany (1953)
251. Union Chemique Belge. Process for making the steel surface hard and corrosion resistant. Patent FR 1042076A, Union Chemique Belge SA, France (1953)
252. A. Młynarczak, Structure and properties of galvanic-diffusion Ni-Al coatings produced on carbon steels. Inżynieria Materiałowa **5**, 300–303 (1999). (in Polish)
253. L. Swadźba, A. Maciejny, B. Mendala, W. Supernak, Diffusion layers of MeCrAlY+Al obtained by the Arc-PVD method on nickel alloys. Inżynieria Materiałowa **5**, 308–313 (1999)
254. G. Totten, Heat treating in 2020: what are the most critical issues and - what will the future look like? Heat Treat. Metals **31**(1), 1–3 (2004)
255. K. Funatani, Heat Treating R&D in Japan – status and challenges, in *Proceedings of the ASM Heat Treating Society's 21st Conference, Nov 5–8, 2001* (ASM International, Indianapolis, 2001), pp. 432–440

256. S. Segerberg, Heat treating R&D in Sweden – status and challenges, in *Proceedings of the ASM Heat Treating Society's 21st Conference, Nov 5–8, 2001* (Indianapolis, 2001), pp. 441–445
257. R. Reynoldson, Heat treating R&D in Australia – status and challenges, in *Proceedings of the ASM Heat Treating Society's 21st Conference, Nov 5–8, 2001* (Indianapolis, 2001), pp. 448–457
258. American Society for Metals. Heat treating technology roadmap update. Vision 2020: looking ahead. Heat Treat. Prog. **1–2**(1–2), 2007 (2007)
259. T. Inoue, K. Funatani, G. Totten, Process modeling for heat treatment: current status and future developments. J. Shanghai Jiaotong Univ. **E-5**(1), 14–25 (2000)
260. R. Houghton, Heat treating technology roadmap update 2006. Part I: process & materials technology. Heat Treat. Prog. **5–6**, 1–4 (2006)
261. G. Pfaffmann, Heat treating technology roadmap update 2006. Part III: equipment and hardware materials technology. Heat Treat. Prog. **11–12**, 1–4 (2006)
262. S. Sikirica, D. Welling, Heat treating technology roadmap update 2006. Part II: energy and environmental technology. Heat Treat. Prog. **8–9**, 1–4 (2006)
263. R. Fabian, An R&D plan for the Heat Treating Community. Technical report, ASM International, Materials Park (1999)
264. R. Gaster, ASM Heat Treating Society's 1999 research & development plan. Report, ASM International, Materials Park (1999)

Chapter 3
Development of Carburising Towards Agile Production

No problem can withstand the assault of sustained thinking.

François-Marie Voltaire, 1694–1778

3.1 Unit, Volume and Mass Production

The literature distinguishes between three basic types of production: unit, volume and mass production. Mass production consists in producing a large number of identical products according to the same technology; and although it is characterised by a very small range, its undeniable advantage over the other types is its low production costs. Volume production (small-, medium- or large-volume production), smaller than mass production, involves the manufacturing of products in batches (or lots, customarily up to 5,000 units) and economically falls between mass production and unit production.

The birth date of mass production is considered to be 1 December 1913, when the Ford Motor Company officially started the first production line for the assembly of the Ford T car. The introduction of a treadmill moving in the floor of the shop reduced the time it took to assemble the car from 12 h to 1.5 h and reduced its price from $800 to $300. In 1914, more than 300,000 pieces of T model rolled out of the assembly plant, making 90% of the cars moving around the world at the time being Fords. Henry Ford was the first to fully understand the importance of standardisation (repeatability), without which mass production is not viable [1].[1] As can be seen, the specialisation of production allows to achieve better quality and quantity results, thanks to the repetitive nature of the work and the continuous improvement of a well-defined production system.

[1] Ford T was the first car that could be repaired with generally available spare parts.

E. Wołowiec-Korecka, *Carburising and Nitriding of Iron Alloys*, Physical Chemistry in Action, https://doi.org/10.1007/978-3-031-59862-3_3

It should be emphasised that important factors in high-volume and mass production are the accuracy and repeatability of the products fabricated. The accuracy is meant as the degree of conformity of the obtained result of a single measurement of a product characteristic with the expected value, while repeatability is meant as the precision of the results obtained under the same measurement conditions. This means that in mass production, all components are expected to have properties with values in line with the expected values and the smallest possible deviation of these values from the average value in the series. Statistical issues have hence an important place in mass production. There is also the concept of process ability (i.e. obtaining the same properties at the same point for all elements in a series) and statistical process control (SPC).

Every production process is influenced by various time-varying factors affecting the repeatability of the manufactured products, so it is important to assess the impact of these factors on the quality of the process. The degree to which the process parameter values are kept within the natural range of the process parameter can be determined by the value of the process parameter scatter coefficient C_p and by the position coefficient of the mean value in relation to the tolerance centre value C_{pk}. Both these indicators characterise the qualitative ability of the production process [2].

$$C_p = \frac{\text{USL-LSL}}{6\sigma} \tag{3.1}$$

$$C_{pk} = \min\left\{\frac{\text{USL} - \bar{x}}{3\sigma}; \frac{\bar{x} - \text{LSL}}{3\sigma}\right\} \tag{3.2}$$

where: USL—upper specification limits (upper tolerance for the property under examination), LSL—lower specification limits (lower tolerance for the property under examination), σ—standard deviation, $\overline{x}$—average value for the property under examination.

The indicator C_p described by Eq. (3.1) specifies how many times the interval of the actual variation of the process parameter falls within the natural range of the process parameter as determined by the value 6σ ($\pm 3\sigma$). The process parameter distribution position coefficient C_{pk} (Eq. 3.2) is a measure of the distance between the actual mean value of the distribution ($\overline{x}$) and the centre of the tolerance interval. On their basis, the qualitative ability of the production process is diagnosed (Table 3.1).

Although the worldwide standard is to assume the values of C_p and C_{pk} at the level of 1.33, which implies a defectivity of 0.0063% of the manufactured items, the ideal situation occurs when both indicators have a value greater than 1.66—only a single defective item is expected in a million products.

The issue raised is of relevance to surface hardening, i.e. carburising with subsequent quenching, used massively and in large volumes in the automotive industry. The aim of this technology is to obtain repeatable, precisely planned parameters for the carburised layer and to minimise negative phenomena affecting dimensional changes after quenching. It is estimated that the current cost to correct post-heat

Table 3.1 Diagnosing the qualitative ability of a production process. Reprinted with permission from Ref. [2] Open access under a CC BY 4.0 license, https://creativecommons.org/licenses/by/4.0/

C_p coeff. value	C_{pk} coeff. value	Diagnosing the qualitative ability of a production process
$C_p = 1$	$C_{pk} = 1$	Production process qualitatively efficient. Expected fraction of defective products: 0.27%
$C_p < 1$	$C_{pk} > 1$	Production process qualitatively inefficient. Requires improvement or widening of tolerances
$C_p = 1.33$	$C_{pk} = 1.33$	Production process with large qualitative ability. Expected fraction of defective products: 0.0063%
$C_p = 1.66$	$C_{pk} < 1$	Production process qualitatively inefficient. The process is influenced by a systematic factor that needs to be removed
$C_p > 1.66$	$C_{pk} > 1.66$	Ideal qualitative ability of the production process. Expected fraction of defective products at the level of *ppm*: 0.0006%

treatment deformation in the worldwide automotive industry (passenger car gearbox industry and in the bearings industry) is roughly 20 billion Euro per year [3].

3.2 Atmospheric Carburising in Mass Industry

The gas carburising process is carried out in the temperature range of 880–950 °C (most commonly at 920 °C), in an atmosphere generated in the furnace by the combustion of hydrocarbons or from liquid organic compounds (methanol, paraffin). The method of carburising in an endothermic atmosphere is most commonly used in the high volume production of parts for the automotive industry, as well as aerospace and machine industry [4]. It is estimated that 90% of the world's gears are processed specifically by atmospheric carburising [5]. In small- and medium-volume production, gas carburising is also carried out in an atmosphere produced "*in situ*".

Gas carburising processes are equilibrium processes,[2] where carbon can be extracted from the following reactions:

$$2CO \rightarrow C + CO_2 \quad \text{(Boudouard's reaction)} \tag{3.3}$$

$$CH_4 + O_2 \rightarrow C + 2H_2O \quad \text{(methane dissociation reaction)} \tag{3.4}$$

$$CO + H_2 \rightarrow C + H_2O \quad \text{(water gas reaction)} \tag{3.5}$$

These reactions occur simultaneously, but at different rates. The water gas reaction occurs fastest, so that despite the presence of small amounts of other carbon-bearing constituents (e.g. methane) in the atmosphere, they are treated as inert compounds, not involved in the process [4].

In the process of carburising in an endothermic atmosphere, the treatment atmosphere is generated in endothermic generators by burning fuel gas (natural gas, butane and propane) with an oxygen deficiency. An equilibrium is created between the resulting atmospheric constituents and a certain proportion of the gases CO/CO_2 and H_2O/H_2 is formed, which also depend on temperature and pressure. This, in turn, determines the ability of the atmosphere to carburise to a certain carbon concentration, i.e. the so-called carbon potential. Maintaining the correct carbon potential during the process is the main prerequisite for the correct implementation of carburising and the formation of the correct structure of the carburised layer. The value of the carbon potential is monitored during carburising most often by means of an oxygen probe (less often a dew point sensor or by determining the CO_2 content) [6].

In the process of carburising in an atmosphere produced "*in situ*", the treatment atmosphere can be produced by dosing liquid organic compounds or organic compounds and water directly into the furnace chamber. These compounds undergo thermal decomposition (pyrolysis) when they enter the hot furnace chamber. The chemical composition of the atmosphere and its properties depend on the organic compounds used. Methods based on the decomposition of organic compounds use methanol, paraffin or a combination of isopropyl alcohol and water, amongst others [4].

As mentioned earlier, the method of gas carburising in an endothermic atmosphere has found widespread use in high volume production [5, 7] due to its ability to produce a large amount of treatment atmosphere. The undoubted advantages of this process are its simplicity, full control of the carbon potential and obtaining a relatively accurate expected carbon profile. However, carbon transfer in an endothermic atmosphere does not have a high efficiency (only a few tens of g/m^3), resulting in the need to

[2] Equilibrium process means that an equilibrium state $w\text{A}+x\text{B} \Longleftrightarrow y\text{C}+z\text{D}$ is established during the process between the substrates and products of the chemical reaction taking place in the process, which can be described with the equilibrium constant of the reaction $K = \frac{[C]^y \cdot [D]^z}{[A]^w \cdot [B]^x}$, where [A], [B], [C], [D]—concentrations of reactants (products and substrates) [mol/dm^3], w, x, y, z—stoichiometric coefficients in the reaction equation. In non-equilibrium processes, the equilibrium of the reaction is disturbed by external stimuli, e.g. a change in substrate concentration, a change in pressure or temperature, making it impossible to determine the equilibrium constant of the reaction.

supply a large amount of atmosphere to the furnace. Due to the presence of oxygen in the treatment atmosphere, the internal oxidation that occurs near the surface of the treated workpieces cannot be eliminated either, resulting in the need to grind the surfaces of the products after the process. Temperature limitations associated with the design of the equipment itself also exist.

3.3 Low-Pressure Carburising

A scientific discussion on whether it is possible to use a reduced pressure in carburising arose in the 1960s and the impetus to start research into carburising with a controlled supply of carbon was the high price of gas used in gas carburising in the 1970s [8]. Low-pressure carburising (LPC), commonly referred to as vacuum carburising, was developed in late 1968 and subsequently patented by H. Westeren (CI Hayes, USA). The process was commercialised in early 1969, but, as D. Herring notes, it took almost three decades of work by researchers and practitioners from around the world to fully accept the process. Development efforts embrace areas such as improvements in design and construction of vacuum furnaces, development of low-pressure carburising methods, optimisation of the process for carbon-bearing gases, creation of empirical databases and construction of process simulators, development of high-pressure gas quenching technology and optimisation of oil quenching techniques, increase in the availability of low-cost alloys for vacuum carburising and gas quenching, etc. [9].

Elimination of oxygen

The first low-pressure carburising equipment used methane and propane under reduced pressure, with both, gas type and pressure influencing the final advantages of low-pressure carburising over atmospheric carburising (endothermic carburising). The essence of the advantage is explained by the chemistry of the process and the phenomenon of chemical adsorption between gases and metals. Adsorption is a surface phenomenon in which substances such as gas, liquids or dissolved solids are trapped on the surface of another substance (here a metal) by the force of a chemical bond.[3] The adsorption phenomenon of the gases used in carburising occurs at different intensities for different metals, for some metals not occurring at all.

Knowing the order of intensification of the adsorption of gases used in gas and vacuum carburising, it is easy to explain the phenomenon of surface oxidation of iron alloys commonly occurring in gas carburising:

$$O_2 > C_2H_2 > C_2H_4 > C_3H_6 > C_3H_8 > CO > H_2 > CO_2 > N_2 \qquad (3.6)$$

---

[3] Another phenomenon is absorption, i.e. the process whereby a substance is absorbed by the entire volume of another substance.

Table 3.2 Hydrocarbon combinations applied for low-pressure carburising

	Mixtures used in the LPC process
Based on acetylene (C_2H_2)	C_2H_2, aC_2H_2/bN_2, aC_2H_2/bH_2, $aC_2H_2/bC_2H_4/cH_2$
Based on cyclohexane (C_6H_{12})	C_6H_{12}, aC_6H_{12}/bC_2H_2
Based on methane (CH_4)	CH_4, aCH_4/bC_3H_8
Based on propane (C_3H_8)	C_3H_8, aC_3H_8/bH_2, aC_3H_8/bN_2

a, *b*, *c*—coefficients, which denote the percentages of the individual gases

According to the relationship given in (3.6), oxygen is an extremely active gas that binds chemically to most metals, while nitrogen is an inert gas with virtually no adsorption on the surface of metals. When a strongly adsorbed gas coexists with a low-adsorption gas in a process chamber, only the strongly adsorbed gas can adsorb on metals, while the weakly adsorbed gas cannot. The amount of chemically adsorbed gas depends on the partial pressure of the gas, which determines the state of saturated adsorption. Once a certain partial pressure is exceeded, no further gas adsorption occurs. A highly adsorbed gas reaches a saturated state at low pressure, while a low-adsorption gas only reaches a saturated state at high pressure. As T. Watanabe and T. Hirata explain, when the steel material is placed in the carburising chamber, the gas with the highest absorption is adsorbed first. If the carburising atmosphere contains oxygen at a sufficiently high partial pressure, the entire surface of the steel material will be covered with oxygen until a saturated state is reached in which no other gases can adhere to the steel material. At a lower partial pressure of oxygen, which prevents the entire surface of the steel material from being covered, a second adsorption gas (such as acetylene) can adsorb into the steel material through gaps in the layer of adsorbed oxygen molecules. The adsorbed acetylene dissociates under the catalytic action of iron, becoming a source of carbon that penetrates deep through the surface of the steel material. In the case of low-pressure carburising, in which no oxygen-containing atmospheres are used (Table 3.2), the undesired phenomenon of oxidation of the carburised material at the grain boundaries is eliminated [10].

Limitation of soot-forming phenomena

Numerous researchers reported in the 1970s and 1980s that the vacuum carburising process was proceeding with avalanche-like rapidity, making it difficult to control. The main problems of excessive carburising were the large amount of cementite released in the carburised layers and soot deposited on the chamber components of the carburising equipment. Excessive amounts of cementite were difficult to dissolve and diffuse, while release of soot substantially increased the operating costs of the equipment. T. Watanabe and T. Hirata explained such highly concentrated carburising by the properties of the carburising gases used at the time, which achieved 100% adsorption of saturation (e.g. methane). In the case of methane, its decomposition reaction under low-pressure carburising conditions ((3.7) reaction) proceeds to the right until a constant partial pressure ratio CH_4/H_2 is reached, as carburising is performed at zero partial pressure of hydrogen (soot is formed by the decomposition of hydrocarbons due to the lack of hydrogen). When a certain amount of hydrogen is present in the carrier gas for carburising, the reaction to the right is not induced as long as the partial ratio of hydrogen to methane is sufficiently high.

$$\mathrm{CH_4} \rightarrow C_{(\mathrm{soot})} + 2H_2 \tag{3.7}$$

For practitioners of the time this meant that the problem of excessive carburising could be avoided by keeping the partial pressure of the coal-bearing gas low, so that the carburising gas did not achieve high adsorption on the metal surface (low-pressure, constant-flow carburising). Alternatively, as soon as high adsorption was achieved, the gas supply to the chamber had to be closed, creating a withstand (annealing) stage in which the carburising gas was being exhausted. The cycle could then be repeated. This solution has been given the name of impulse carburising [10].

Aliphatic hydrocarbons

The general progress of soot-forming hydrocarbon decomposition is expressed by the Eq. (3.8). The progress intensity of this reaction depends on the strength of the bonds between the carbon and hydrogen atoms in its molecule (Table 3.3). In the case of a chemical compound with a high-carbon-carbon bonding strength, the reaction (3.8) does not proceed to the right, even if the ratio of partial pressures H_2/C_nH_m is small. In fact, this means that acetylene is the most effective gas for soot-free carburising. It has a high adsorption strength to iron and, moreover, even at low hydrogen content in the atmosphere, it does not decompose without the participation of the catalytic surface. The triple-bonded hydrocarbon does not readily decompose thermally due to the strong bonding strength of carbon in the molecules, but the high adsorption strength to iron means that it readily adsorbs to the carburised surface without stopping the build-up of the carburised layer. Obviously, the absence of soot on carburised surfaces is strongly dependent on the carburising temperature; at lower temperatures, catalytic decomposition processes proceed faster than the diffusion process, causing an oversupply of carbon to the surface.

$$C_nH_m \rightarrow nC_{(\mathrm{soot})} + \frac{m}{2}H_2 \tag{3.8}$$

The next step in the development of low-pressure carburising was the optimisation of the dosing of carbon-bearing gases to the surface, depending on temperature and saturation time, which was possible once computer calculations came into general use. This process is described in the Sect. 6.5.

Maturity of low-pressure carburising

At present, the carburising process is quite well understood, both in terms of its kinetics [12–14], the influence of individual factors [15, 16], surface phenomena [17], as well as the final properties of the carburised material [18–23].

Table 3.3 Bonding energy of methane, ethane, ethylene and acetylene [10, 11]

Hydrocarbon	Bond distance [Å]		Bond energy C-C [kcal/mol]
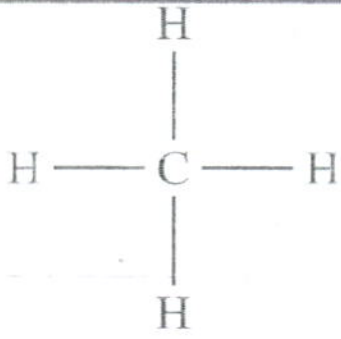 Methane	–	1.087	–
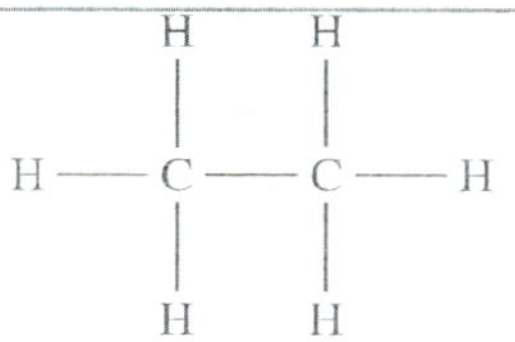 Etan	1.535	1.094	83
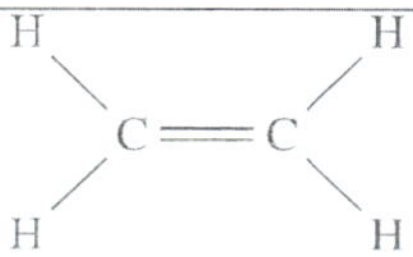 Ethen (ethylene)	1.339	1.087	145
 Ethyne (acetylene)	1.203	1.060	198

In industrial practice, low-pressure carburising operates in two varieties, based on two historically different concepts adopted in the 1970s. The first is carburising at low pressure and with a constant flow of gas for the entire duration of the process and is only feasible in a dedicated vacuum device. The second variation is vacuum carburising by pulsed gas injection, which can even be implemented in a device with a design based on a gas carburising device. For continuous carburising, it is difficult to ensure an equal supply of carbon-bearing gas to the surface and interior of the charge, leading to a large spread in the thickness of the resulting layers within the charge. Pulsed acetylene carburising has been shown to be effective in achieving homogeneity of products in a batch, as the introduction of acetylene with a certain excess ensures absolutely homogeneous adsorption saturation (the saturated state of gas adsorption on metals is limited to one molecular layer and the excess ensures that there are no discontinuities in adsorption). A certain disadvantage of this method is that it consumes more carbon-bearing gas than is necessary [10].

The *FineCarb* technology being in use in the industry is carburising of the second type. It is a complex process in which small amounts of carbon-bearing gas are cyclically admitted and then removed under low pressure (1–10 hPa) at temperatures in the range of 900–1,050C °C, at a rate that allows carbon on the steel surface to reach the limit (maximum) solubility of carbon in austenite [24–27]. At the stage of saturation (the stage where carbon-bearing gases are injected into the chamber), an intensive transfer of carbon occurs from the low-pressure atmosphere to the steel surface, causing a rapid growth in the surface concentration of carbon. In the withstand stage, the gas supply is stopped and the component is annealed at an additionally reduced pressure, causing carbon to diffuse deep into the material and causing a drop in concentration at the surface. The transfer of mass during carburising takes place in three stages: (1) transport of carbon from the atmosphere to the surface of the steel, (2) the chemical reaction at the gas-steel interface and (3) the diffusion of carbon deep into the material. Carbon diffusion between the carbon-rich atmosphere and the steel surface is the driving force behind mass transfer phenomena and sets the carbon concentration gradient in the surface layer of the component being carburised [28]. In this variant of carburising, carbon for diffusion saturation of austenite is obtained by thermal decomposition of short-chain (aliphatic) carbon-bearing gases (acetylene and ethane), in the manner described by W. Grafen, B. Edenhofer and P. Kula [4, 29].

Compared to atmospheric processes, low-pressure treatment has many advantages (cf. Table 3.4). Carburising conducted in a reduced-pressure environment protects the material from oxidation across grain boundaries ensuring high surface quality [24, 30–32]. As a consequence of high process temperatures, carbon transfer is higher than in the conventional carburising method, making low-pressure carburising a more efficient process compared to endothermic carburising (allowing to shorten the process time). Moreover, LPC does not pollute the environment and uses less energy. In combination with quenching in inert gas, it causes the minimum deformation of the components [24, 30, 32, 33]. For this reason, LPC in combination with high-pressure gas quenching (HPGQ) is commonly used, inter alia, for the hardening of transmission components of drive systems in the automotive industry [34, 35].

In terms of mechanical properties, each variation of the carburising process gives a hard, wear-resistant surface and, at the same time, the ability to transfer high stresses due to the flexible core. These characteristics are sufficient to satisfy most of the needs of industries that employ carburising to improve the quality of steel components [22, 34, 36].

At the beginning of the 2000s, low-pressure carburising, developed at the end of the 1960s [29, 37–40] and by the end of the twentieth century fully commercialised [41–48], is seen as a technology with great potential for development. The scientific and technological state of low-pressure carburising at the dawn of the twenty-first century meets most of the then commonly expected requirements, while leaving a great field of activity for research into the optimisation and efficiency of processes.

In the scientific field, an in-depth understanding of the transients of low-pressure processes, their boundary conditions and interactions at the atmosphere-material interface has been sought over the past two decades [13, 24, 49–53]. Much attention

Table 3.4 Comparison of gas and low-pressure carburisings [4, 10]

Comparison element	Atmospheric carburising	Low-pressure carburising
The treatment temperature	Maximum process temperature is limited by the equipment design	The maximum process temperature can be 100–150 °C higher than that in atmospheric carburising
Treatment time	Reducing the carburising time by increasing the temperature is not possible. A constant working time of the atmosphere is needed	It is possible to reduce the treatment time by high-temperature carburising, a constant time during which the charge stays in the atmosphere is not necessary
The carbon surface concentration	Carbon surface concentration is controlled by the carbon potential of the atmosphere	Carbon surface concentration depends on the decomposition process of hydrocarbons and is usually very high
Oxidation of grains boundary	Oxidation at the grains boundary occurs at the surface of the carburising layer, thus lowering the strength of the workpiece	As the operation occurs at a pressure close to vacuum, oxidation of the grain boundaries does not occur
Gas consumption	Methanol or endothermic gas is largely consumed as carrier gas	As carrier gas is not needed, less gas is generally consumed
The rate of diffusion saturation	Rate of diffusion saturation is small and is limited by the temperature and carbon potential of the atmosphere	Rate of diffusion saturation can be high due to higher process temperature and sudden breakdown of carbon-bearing hydrocarbons
Greenhouse gas emissions (CO_2)	During the process, 2–5 kg of CO_2/hour is discharged from the device	Small amounts of decomposition products of carbon-bearing gases (e.g. C_2H_2) are emitted. CO_2 is not released into the atmosphere
The stability of carburising quality	Changing the composition of the atmospheric gas by mixing with air influences the quality of carburising	A small dispersion of carburising conditions, stable quality and repeatable results are obtained
Operability	Intermittent working significantly reduces productivity	Intermittent working is possible without reducing productivity
Batch cooling	Parts are cooled in oil and need to be washed	Parts can be cooled in oil or gas. Gas cooling does not require washing the parts
Thermal deformations during treatment	There is a limit in terms of conventional deformations	Low distortions are made possible by applying gas quenching
Working environment	Unfriendly working environment (fire curtains, presence of spent gas, etc.)	No fire curtain or exhaust fumes or good working environment
Cost of equipment and accessories	Small	Vacuum-related apparatuses are necessary, and their price is significant

has been paid to carbon-bearing atmospheres, with a particular focus on acetylene in reduced pressure [54–62]. Most of the research in this area was done with the support of computer-based computing tools. Between 2001 and 2023, computer techniques have become a common working tool for predictive studies to predict the effects of the progress of carburising processes and this is a worldwide trend connected with the technological and financial availability of computers' computing power [63–68].

A continuous reduction in machining costs by speeding it up was expected in the technological and industrial areas. Such an effect has been seen in the development of high-temperature carburising processes (above 1,010 °C), steels for such carburising (resistant to grain growth above 1,010 °C) [17, 32, 69–76] and effective carburising models integrated with process control systems for the purpose of real-time process control (cf. Chap. 6). This has intensified works in these areas, as well as in the area of austenitic carbonitriding, which allows to considerably enhance the strength properties of cheap steels [77–84]. Research over high-temperature low-pressure austenitic carbonitriding was conducted by P. Kula, M. Krasowski and S. Pawęta between 2006 and 2014. They proved that feeding ammonia at the initial stage of the process (the stage of heating the elements for carburising) and introducing nitrogen into the surface layer at this stage have a favourable effect on the growth kinetics of the carburised layer (Table 3.5) and reduces the growth of grains (Fig. 3.1) [17, 85–87].

A number of scientific reports from 2001 to 2022 also pertain to investigations into the application of carburising of steels with special chemical compositions, such as high-alloy steel Ferrium C61, CSB-50NIL, Incoloy 800HT, Pyrowear 675, Ferrium 69, chromium-nickel steels [21, 35, 88–93] or for specific components subject to mass production, such as helical gears, injector flanges in diesel engines and automotive gear rings [94–96].

In view of the expectations raised by the industry to obtain more repeatable and homogeneous carburised layers, they have set a new direction for research into low-pressure carburising.

Table 3.5 The influence of carburising temperature according to *PreNit*LPC on treatment time (steel EN 16MnCr5). Reprinted with permission from Ref. [97] Copyright (2011) ASM International

CD_{Eff} [mm]	Carburising time [h:min]					
	920 °C	950 °C	980 °C	1,000 °C	1,020 °C	1,040 °C
0.5	1:23	0:57	0:39	0:30	0:24	0:19
1.0	5:30	3:50	2:35	2:00	1:35	1:15
2.0	22:00	15:10	10:20	8:00	6:10	4:50
	100%	69%	47%	36%	28%	22%

CD_{Eff}—the thickness of the effective carburised layer, determined with the criterion of carbon concentration 0.35%C

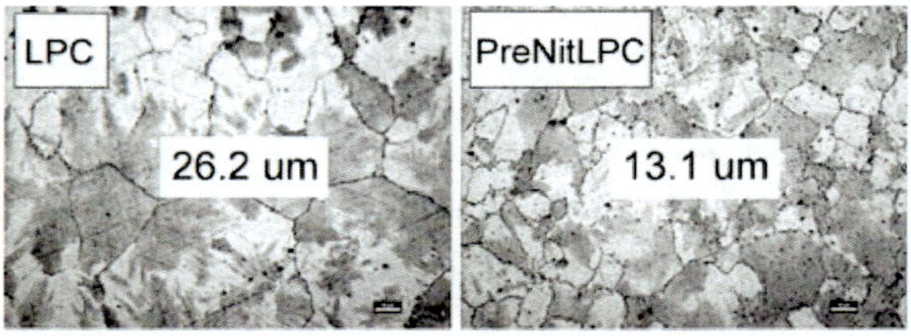

Fig. 3.1 The comparison of grain size after the LPC and *PreNit*LPC treatment at 1000 °C for steel EN 14NiCr6. Reprinted with permission from Ref. [97] Copyright (2011) ASM International

3.4 Low-Pressure Single-Piece Flow Carburising

In low- and medium-volume production, the surface hardening process (*case hardening*) is carried out in single-chamber furnaces with an integrated high-pressure gas quenching system or in double-chamber furnaces with a separate gas or oil cooling chamber. For the surface hardening of power transmission components, the traditional technology, implemented in atmospheric equipment with oil quenching, is predominant globally, accounting for about 80–85% of production. The remaining 15–20% belongs to modern vacuum technology, the area of which is gradually expanding, at the expense of traditional treatments.

As regards productivity and process costs, atmospheric pass-through devices (e.g. pusher furnaces) have a clear advantage, but at the same time are scored the lowest in terms of quality, production organisation and flexibility, monitoring, control and occupational safety, with such issues being increasingly important in modern industry [98] (see Table 3.6). For this reason, vacuum devices in the form of modular systems or three-chamber furnaces are being installed more and more often in place of atmospheric furnaces. Both solutions are similarly evaluated and are used alternatively, effectively competing with atmospheric devices [98].

A series of workpieces in the devices described above are processed simultaneously in the process chamber, spreading them out on a tray plane and replicating this spread over several or more tray levels. For mass production, modular systems with multiple process chambers are constructed and a separate loading and unloading chamber serves the process chambers by cooling the components in gas or oil. All of these systems are characterised by the volumetric method of heating and cooling of the batches, so it is difficult to guarantee identical, completely reproducible process conditions for each individual component [99].

In summary, homogeneity and repeatability of hardened layers are at the heart of solving the problem of manufacturing precision. Physical homogeneity of a solid is understood as exhibiting the identical properties (thermal expansion, electrical conductivity, etc.) in each small volume region of a given substance on a macroscopic scale. Homogeneous hardened layers exhibit the same properties irrespective of the body point at a certain distance from the surface where the given property is determined. Repeatability means that the independent results for the determination of the same material, obtained using the same analytical method, are consistent under identical conditions. In the case of hardened layers, this means that

Table 3.6 Comparison of manufacturing-operating parameters of commercially available heat and thermo-chemical treatment installations

Equipment and Criterion	Atmospheric pusher furnace	Vacuum modular system	Triple-chamber vacuum furnace	Single-piece flow furnace
Investment cost	++	+	+	o
Productivity	+++	++	+	o
Process economics	++	+	+	o
Precision and repeatability of results	o	+	+	+++
Reduction and control of deformation, Reduction of machining costs	o	+	+	+++
Integration in the production line, reduction of logistics costs	o	+	+	+++
Compactness of the installation	o	+	+	++
Operational flexibility	o	+	++	+
Process and online quality control	o	+	+	+++
Safety	o	+	+	+
Environmental protection	o	+	+	+

+/++/+++—degree of advantage of the device over other types
o—no advantages

successive manufactured components examined using the same research method, under identical conditions, will have consistent values for the examined characteristics such as hardness or carburised layer thickness (within the accuracy to the measurement error of the test instrument). This means that the best possible solution to the issue is to provide all components in the charge with repeatable treatment conditions.

When considering to ensure identical process conditions for each workpiece in the series, the concept comes to mind of treating workpieces one at a time in a chamber, with the identical chamber location and identical process conditions in the chamber. For carburising in batch furnace, which takes dozens of minutes, to several hours, achieving repeatability in this way would significantly worsen the economics of the process. Treatment with the continuous feeding of individual workpieces, in a process is a more efficient solution, yet one that retains the most important

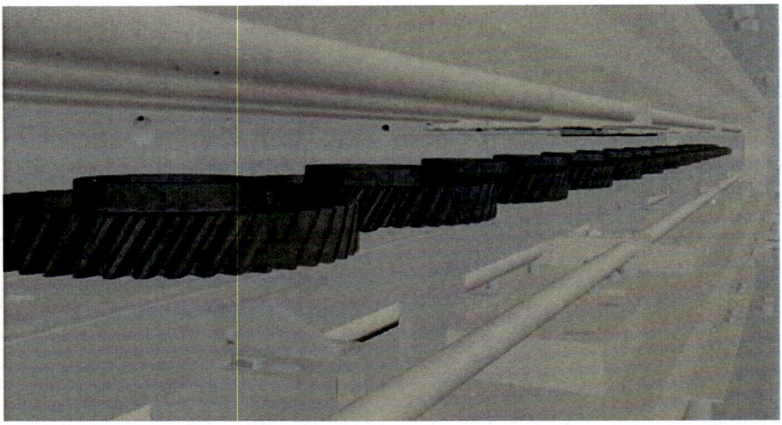

Fig. 3.2 Low-pressure carburising chamber with continuous feeding of individual workpieces (single-piece flow carburising method). Reproduced with permission of Seco/Warwick

features of the previous solution. Such a solution is described in M. Korecki's works [99, 100], where the carburised workpieces pass through dozens of positions during treatment, where all components receive identical process conditions (single-piece flow carburising method) (Fig. 3.2). A vacuum carburising device with a continuous feeding of individual workpieces consists of three process chambers for heating, saturation and annealing, respectively. The carburised components pass through all positions in 3 chambers in turn: in the heating chamber they are heated to process temperature, then in the saturation chamber the workpieces are subject to vacuum carburising by introducing carbon-bearing gas. Further on, the workpieces go into the annealing chamber, where they pass through positions with the carburising process temperature and then through positions with subcooling temperatures for quenching. Finally, the workpieces are ready to be quenched and tempered [101].

The results of the studies published in 2018–2022 prove that vacuum carburising carried out with the method with a flow movement of individual components provides carburised layers with an accuracy in the range of the measurement error of the layer thickness [100, 101]. The presented studies of microstructures also confirmed the correctness of the obtained carburised layers produced by the flow method. It was demonstrated that the layers are homogeneous (in the sense that they are uniform within an element) and repeatable (in the sense they can be reproduced on any number of elements) to an accuracy of 4%. It was estimated that the flow method provides the repeatability of layers at least five times better than those produced in modern thermo-chemical treatment equipment (up to ten times better in the case of ranges permitted by some customers). Thus, carburised layers produced by the flow method are suitable for the production of precision machine parts where there are high restrictions on the properties of the parts after machining, as is the case in the automotive industry.

The variation coefficient V presented in Table 3.7 is a measure of the variation in the distribution of a characteristic, allowing to assess the variability of the series independent of the size of the arithmetic mean, calculated as $V = \frac{s}{\overline{x}} \cdot 100\%$, where s—the standard deviation from the sample, and $\overline{x}$—the arithmetic mean from the sample. In Table 3.7, comparing the repeatability of the carburised layers produced in an example of the low-pressure carburising process and the single-piece flow low-pressure carburising process, it was used to characterise the measure of variation in the experimentally determined carburised layer' thicknesses (CD_{Eff}) in a single-chamber and flow device, analysed for repeatability. The variation coefficient for the flow method V is 6 times smaller than for carburising in the charge.

The potential ability of the flow process shown in Table 3.7 (i.e. parameter scatter coefficient CD_{Eff} of the process C_p) is 2.08, and the position coefficient of the mean value in relation to the tolerance centre value C_{pk} is 2.04. A value of both coefficients above 1.66 indicates a very good qualitative ability of such a production process and the expected share of defective products does not exceed 0.000006 (cf. Table 3.1) [102].[4]

A several-fold improvement in the repeatability of the process results (uniformity and repeatability of the carburised layer thickness in a single component and in the whole series) comes from the fact that each workpiece receives identical process parameters: temperature and atmosphere. A workpiece heated individually heats faster and more uniformly, and the effect of temperature non-uniformity in the process chamber volume, which could occur at different positions, is eliminated by averaging as a result of the component passing through all positions.[5] It is similar with the process atmosphere, which directly reaches the surface of the component and, even if its composition changes at different positions, the average value, after the components have passed through all positions, is the same.

To recapitulate, the method of low-pressure carburising with the continuous feeding of individual workpieces (low-pressure single-piece flow carburising), produces results with less variability than the method of atmospheric and batch low-pressure carburising. It should be however noted that the characteristics of this process: reduced carbon supply, indefinite initial conditions, the need to individually select the temperature and process cycles to suit the expected final profile, make it impossible to implement the process without a control system based on fast numerical calculations. This aspect fits in with the idea of Society 5.0 and will be discussed further in Chap. 6.

[4] The lower (LSL) and upper (USL) specification limit were taken to be 0.50 and 0.60 mm, respectively, as the typical range of required thicknesses of carburised layers.

[5] Studies to date confirm that a temperature uniformity effect is reached of less than 1 °C.

Table 3.7 Comparison of the repeatability of carburised layers produced by the example low-pressure carburising process and by the low-pressure single-piece flow carburising process

Lp	CD_{Eff} [mm]	
	Low-pressure carburising	Low-pressure single-piece flow carburising
1	0.78	0.55
2	0.76	0.55
3	0.85	0.54
4	0.78	0.55
5	0.87	0.54
6	0.89	0.56
7	0.85	0.56
8	0.81	0.56
9	0.79	0.55
Mean CD_{Eff} ($\overline{x}$, mm)	0.820	0.551
Std deviation (s, mm)	0.046	0.008
Range (r, mm)	0.130	0.020
Variation coeff. (V)	**6%**	**1%**

CD_{Eff}—thickness of the effective carburised layer, determined with a carbon concentration criterion of 0.4%C, with a precision of 0.01 mm

3.5 Further Directions of Development

Carburising in endothermic atmospheres is a process that has been known for decades and is broadly used. However, its nature entails a serious disadvantage, which is oxidation across austenite grain boundaries (IGO), caused by the presence of oxidising gases (carbon dioxide, water) in the atmosphere. This necessitates the need to grind the surface to remove dangerous structures, even to a depth of tens of micrometres, which generates high costs [7]. Carbon transfer in an endothermic atmosphere does not have a high efficiency, resulting in the need to supply a large amount of atmosphere to the furnace. The costs relate not only to the significant quantities of agent consumed (typically methane, propane or methanol), but also to energy losses and greenhouse gas and energy emissions. The flammability and explosiveness of the furnace atmosphere and the associated hazards are not insignificant for the environment, and strict safety and environmental regulations must be applied [103].

It is estimated that there are 200,000 carburising plants operating in the global industry, of which 75–80% are atmospheric plants and 15–20% vacuum plants. A gradual process of replacing operationally worn-out atmospheric plants with vacuum plants is being also observed, which will increase the market share of low-pressure carburising technologies. This will force a gradual shift away from this technology, just as global trends are moving away from sources of fossil-based raw materials and energy (natural gas, oils).

As for the optimisation of the carburising process, it seems that further works are only possible by optimising numerical calculations, as is the case with low-pressure carburising technology, where an efficient and effective method of process utilisation (exposure time to saturation, flow rate of carbon over time, etc.) is achieved by integrating the technology with computer-based optimisation and simulation systems (see Chap. 6) [26, 92, 104, 105].

References

1. A. Grzeszak, Tape world. Polityka **48**(2935), 48 (2013). (in Polish)
2. J. Szkoda, Diagnosing the states of the qualitative capacity of the production process. Diagnostyka **27**, 89–94 (2002). (in Polish)
3. Institut fur Werkstofftechnic, Economic consequences of distortion. Report, Institut fur Werkstofftechnic, Bremen (1995)
4. P. Kula, *Surface Layer Engineering* (Lodz University of Technology, Lodz, 2000). (in Polish)
5. A. Rakhit, *Heat Treatment of Gears: A Practical Guide for Engineers* (ASM International, Materials Park, 2000)
6. K. Przybyłowicz, *Metallography* (WNT, Warsaw, 2003). (in Polish)
7. G. Parrish, *Carburizing: Microstructures and Properties* (ASM, Materials Park, 1999)
8. C. Luiten, Wärmebehandlung mit Neuartigen Vakuumöfen in den Vereinigten Staaten. Heat Treat. Metals **18**(2), 172–175 (1963). (in German)
9. D. Herring, R. Peters, New-formula acetylene cool for heat treatment. Gear Technol. **9**, 90–96 (2013)
10. T. Watanabe, T. Hirata, New concept and practical operation of carburizing and nitriding. Authors' own edition, Japan (2015)
11. H. Schmalzried, *Solid State Reactions* (PWN, Warsaw, 1978). (in Polish)
12. A. Ochsner, J. Gegner, G. Mishuris, Effect of diffusivity as a function of the method of computation of carbon concentration profiles in steel. Metal Sci. Heat Treat. **46**, 148–151 (2004)
13. R. Gorockiewicz, The kinetics of low-pressure carburizing of alloy steels. Vacuum **86**(4), 448–451 (2011)
14. K. Tanaka, H. Ikehata, H. Takamiya, H. Mizuno, Calculation of microstructure in vacuum carburizing incorporating kinetics modeling of grain-boundary cementite. Iron Steel Inst. Japan **52**, 134–139 (2012)
15. F. Chen, L. Liu, Deep-hole carburization in a vacuum furnace by forced convection gas flow method. Mater. Chem. Phys. **82**, 801–807 (2003)
16. K. Widanka, Effect of phosphorus on vacuum carburising depth of iron compacts. Arch. Civil Mech. Eng. **10**(1), 85–91 (2010)
17. P. Kula, R. Pietrasik, S. Paweta, K. Dybowski, L. Kaczmarek, A. Gładka, High temperature low-pressure carburizing with prenitriding process - the economic option for vacuum carburizing. Chiang Mai J. Sci. **40**(5), 865–873 (2013)
18. A. Rzepkowski, A. Rzepkowski, P. Kula, Vacuum carburized layers examinations of ISO 2639 standard. Acta Metallurgica Slovaca **1**, 730–733 (2004)
19. S. Carey, D. Herring, Low-pressure carburizing process development of M50NiL. Heat Treat. Prog. **5–6**, 43–46 (2007)
20. S. Azis, I. Jauhari, N. Ahamad, Improving surface properties and wear behaviors of duplex stainless steel via pressure carburizing. Surface Coat. Technol. **210**, 142–150 (2012)
21. M. Semenov, Control of heat-resistant steel carburized layer structure. Part I. Metal Sci. Heat Treat. **55**(5–6), 257–264 (2013)

22. M. Korecki, P. Kula, E. Wołowiec, M. Bazel, M. Sut, Low-pressure carburizing and nitriding of fuel injection nozzles. Heat Process. **3**, 59–62 (2014)
23. S. Wei, G. Wang, X. Zhao, X. Zhang, Y. Rong, Experimental study on vacuum carburizing process for low-carbon alloy steel. J. Mater. Eng. Perform. **23**(2), 545–550 (2014)
24. P. Rokicki, K. Dychton, Acetylene flow rate as a crucial parameter of vacuum carburizing process of modern tool steels. Arch. Metall. Mater. **61**(4), 2009–2012 (2016)
25. S. Tsepov, Characteristic features of carburizing of steel during vacuum carburizing. Metal Sci. Heat Treat. **21**(8), 633–638 (1979)
26. P. Kula, R. Pietrasik, K. Dybowski, Vacuum carburizing - process optimization. J. Mater. Process. Technol. **164–165**, 876–881 (2005)
27. S. Tsepov, M. Krishtal, Characteristics of structure formation in the surface layers of metal during vacuum carburizing. Metal Sci. Heat Treat. **25**(5), 358–359 (1983)
28. O. Karabelchtchikova, R. Sisson, Calculation of gas carburizing kinetics from carbon concentration profiles based on direct flux integration. Defect Diffus. Forum **266**, 171–180 (2007)
29. W. Gräfen, B. Edenhofer, Acetylene low-pressure carburizing – a novel and superior carburizing technology. HTM J. Heat Treat. Mater. **26**(4), 79–83 (1999)
30. V. Osterman, T. Jones, Carburizing for our troops. Ind. Heat. **5**, 53–55 (2006)
31. L. Małdziński, M. Korecki, Comparison of the effectiveness of vacuum and gas carburizing, in *Proceedings of 12th Seco/Warwick Seminar: Modern Trends in Heat Treatment, Sept 25–26, 2008*, Bukowy Dworek (2008). (in Poland)
32. P. Kula, K. Dybowski, E. Wołowiec, R. Pietrasik, Boost-diffusion vacuum carburising - process optimisation. Vacuum **99**, 175–179 (2014)
33. K. Loeser, R. Seemann, High-pressure gas quenching and the influence of material hardenability of case hardening steels in the gear industry. Technical report, ALD-Technical Information (2004)
34. D. Kim, Y. Cho, S. Kim, W. Lee, M. Lee, H. Han, A numerical model for vacuum carburization of an automotive gear ring. Metals Mater. Int. **17**, 885–890 (2011)
35. F. Otto, D. Herring, Vacuum carburizing of aerospace and automotive materials. Heat Treat. Progress **1–2**, 33–37 (2005)
36. A. Goncharov, R. Uvarova, Mechanical properties of steels after vacuum and gas carburizing. Metal Sci. Heat Treat. **32**(5), 321–325 (1990)
37. F. Eysell, Uber die Aufkohlung im Unterdruck Bereich, Verfahrensparameter und Anwendung. Elektrowaerme Int. **34**(6), 312–317 (1976). (in German)
38. C. Luiten, F. Limque, F. Bless, Carburizing in vacuum furnaces, in *Proceedings of Heat Treatment'79, May 22–24, 1979*, Birmingham (1979)
39. Y. Pourprix, J. Naudot, Carburizing under reduced pressure. Traitement Thermique **197**, 51–58 (1985). (in French)
40. B. Edenhofer, J. Bouwmann, Progress in design and use of vacuum furnaces with high pressure gas quenching systems. Ind. Heat. **2**, 333–336 (1988)
41. E. Gronquist, V. Scotto, D. Taft, H. Westeren, Method of vacuum carburizing. Patent US 3,796,615, Hayes Inc., USA (1974)
42. V. Krylov, V. Yumatov, V. Kurbatov, Method for cementation of steel parts. Patent USSR 668978, Enterprise, Russia (1979)
43. T. Kubota, Vacuum carburizing method and device, and carburized products. Patent US 5,702,540, JH Corporation, Japan (1997)
44. P. Kula, J. Olejnik, P. Heilman, Method for under-pressure carburizing of steel workpieces. Patent EU 1,558,781, Seco/Warwick, Lodz University of Technology, EU (2006)
45. P. Kula, J. Olejnik, P. Heilman, Hydrocarbon gas mixture for the under pressure carburizing of steel. Patent EU 1,558,780, Seco/Warwick, Lodz University of Technology, EU (2007)
46. R. Poor, G. Barbee, S. Verhoff, J. Brugg, Furnace for vacuum carburizing with unsaturated aromatic hydrocarbons. Patent US 7,267,793, Surface Combustion, Inc., USA (2007)
47. W. Jones, Versatile high velocity integral vacuum furnace. Patent US 7,514,035, Solar Atmospheres Inc., USA (2009)

48. D. Herring, *Vacuum Heat Treatment: Principles, Practices, Applications* (BNP Media II, Troy, 2012)
49. R. Gorockiewicz, A. Łapinski, Structure of the carbon layer deposited on the steel surface after low-pressure carburizing. Vacuum **85**, 429–433 (2010)
50. P. Kula, Ł. Kaczmarek, K. Dybowski, R. Pietrasik, M. Krasowski, Activation of carbon deposit in the process of vacuum carburizing with preliminary nitriding. Vacuum **87**, 26–29 (2013)
51. Y. Yogo, K. Tanaka, In situ observation for abnormal grain coarsening in vacuum-carburizing process. Metall. Mater. Trans. A **45**(6), 2834–2841 (2014)
52. T. Bensabath, H. Monnier, P.-A. Glaude, Detailed kinetic modeling of the formation of toxic polycyclic aromatic hydrocarbons (PAHs) coming from pyrolysis in low-pressure gas carburizing conditions. J. Anal. Appl. Pyrol. **122**, 342–354 (2016)
53. A. Smirnov, M. Ryzhova, M. Semenov, Choice of boundary condition for solving the diffusion problem in simulation of the process of vacuum carburizing. Metal Sci. Heat Treat. **59**(3–4), 237–242 (2017)
54. T. Mendiara, M. Domene, A. Millera, R. Bilbao, M. Alzueta, An experimental study of the soot formed in the pyrolysis of acetylene. J. Anal. Appl. Pyrol. **74**(1–2), 486–493 (2005)
55. R. Khan, S. Bajohr, F. Graf, R. Reimert, Modeling of acetylene pyrolysis under steel vacuum carburizing conditions in a tubular flow reactor. Molecules **12**, 290–296 (2007)
56. R. Khan, *Vacuum gas carburizing - fate of hydrocarbons*. PhD thesis, Universitat Karlsruhe, Karlsruhe (2008)
57. R. Khan, S. Bajohr, D. Buchholz, R. Reimert, H. Minh, K. Norinaga, V. Janardhanan, S. Tischer, O. Deutschmann, Pyrolysis of propane under vacuum carburizing conditions: an experimental and modeling study. J. Anal. Appl. Pyrol. **81**(2), 148–156 (2008)
58. M. Jung, S. Oh, Y.-K. Lee, Predictive model for the carbon concentration profile of vacuum carburized steels with acetylene. Metals Mater. Int. **15**(6), 971–975 (2009)
59. N.E. Sanchez, A. Callejas, A. Millera, R. Bilbao, M.U. Alzueta, Polycyclic aromatic hydrocarbon (PAH) and soot formation in the pyrolysis of acetylene and ethylene: effect of the reaction temperature. Energy Fuels **26**(8), 4823–4829 (2012)
60. K. Yada, O. Watanabe, Reactive flow simulation of vacuum carburizing by acetylene gas. Comput. Fluids **79**, 65–76 (2013)
61. C. Saggese, N. Sanchez, A. Frassoldati, A. Cuoci, T. Faravelli, M. Alzueta, E. Ranzi, Kinetic modeling study of polycyclic aromatic hydrocarbons and soot formation in acetylene pyrolysis. Energy Fuels **28**(2), 1489–1501 (2014)
62. Y. Song, J.-H. Kim, K.-S. Kim, S. Kim, P. Song, Effect of C2H2/H2 gas mixture ratio in direct low-temperature vacuum carburization. Metals **8**(7), 493 (2018)
63. O. Karabelchtchikova, R. Sisson, Carbon diffusion in steels: a numerical analysis based on direct integration of the flux. J. Phase Equilib. Diffus. **26**(6), 598–604 (2006)
64. P. Cavaliere, G. Zavarise, M. Perillo, Modeling of the carburizing and nitriding processes. Comput. Mater. Sci. **46**(1), 26–35 (2009)
65. Y. Wei, L. Zhang, R. Sisson, Modeling of carbon concentration profile development during both atmosphere and low-pressure carburizing processes. J. Mater. Eng. Perform. **22**(7), 1886–1891 (2013)
66. S.-J. Lee, D. Matlock, C. van Tyne, Comparison of two finite element simulation codes used to model the carburizing of steel. Comput. Mater. Sci. **68**, 47–54 (2013)
67. M. Semenov, A. Smirnov, M. Ryzhova, Computation of carbon concentration curves in vacuum carburizing of steels. Metal Sci. Heat Treat. **55**(1–2), 38–42 (2013)
68. L. Zhang, *Modeling and verification of simulation tools for carburizing and carbonitriding*. PhD thesis, Worcester Polytechnic Institute, Worcester (2016)
69. W. Gräfen, B. Edenhofer, New developments in thermo-chemical diffusion processes. Surf. Coat. Technol. **200**(5–6), 1830–1836 (2005)
70. R. Yin, Carburization of 310 stainless steel exposed at 800–1100 °C in 2%CH4/H2 gas mixture. Corros. Sci. **47**(8), 1896–1910 (2005)
71. R. Yin, Cyclic and isothermal exposures of 310SS to 10%CH4/H2 carburizing gas mixture at high temperatures. Mater. Sci. Eng. A **391**(1–2), 19–28 (2005)

72. V. Heuer, G. Schmitt, Ch. Pustowka, High temperature vacuum carburizing in single chamber furnaces. Inżynieria Materiałowa **4**, 41–45 (2009)
73. P. Kula, E. Wołowiec, R. Pietrasik, K. Dybowski, L. Klimek, The precipitation and dissolution of alloy iron carbides in vacuum carburization processes for automotive and aircraft applications – Part I. Adv. Mater. Res. **486**, 297–302 (2012)
74. P. Kula, E. Wołowiec, R. Pietrasik, K. Dybowski, L. Klimek, The precipitation and dissolution of alloy iron carbides in vacuum carburization processes for automotive and aircraft applications - Part II. Adv. Mater. Res. **486**, 303–308 (2012)
75. M. McKimpson, Development in microalloyed, coarsening-resistant steels for high temperature carburizing. Adv. Mater. Process. **9**, 53–55 (2014)
76. S. Hiramatsu, K. Inagaki, H. Sakaue, I. Yamamoto, Study of the high-temperature vacuum carburizing phenomenon and the carburizing prediction model based by big data approach, in *Proceeding of 24th IFHTSE Congress/European Conference on Heat Treatment, June 26–27, 2017*, Nice (2017)
77. H. Wang, W. Wang, Simple model for carbonitride dissolution in microalloyed steels. Mater. Sci. Technol. **23**(7), 833–838 (2007)
78. M. Fares, A. Talhi, K. Chaoui, M. Touhami, Effects of gas carbonitriding process on surface characteristics of new hot working tool steel. Surf. Eng. **27**(8), 595–601 (2011)
79. V. Campagna, R. Bowers, D. Northwood, X. Sun, P. Bauerle, Comparison of carbonitriding and nitrocarburising on size and shape distortion of plain carbon SAE 1010 steel. Surf. Eng. **27**(2), 86–91 (2011)
80. A. Leineweber, T. Gressmann, E. Mittemeijer, Simultaneous control of the nitrogen and carbon activities during nitrocarburising of iron. Surf. Coat. Technol. **206**(11–12), 2780–2791 (2012)
81. K. Kocurova, M. Domankova, M. Hazlinger, The influence of carbonitriding process on microstructure and mechanical properties of micro-alloyed steel. Metalurgija **52**(1), 19–22 (2013)
82. T. Marray, P. Jacquet, M. Mansori, A. Fabre, L. Barrallier, A thermodynamic and experimental study of low-alloy steels after carbonitriding in a low-pressure atmosphere. Metal Sci. Heat Treat. **56**(7–8), 434–439 (2014)
83. X. Gu, G. Michal, F. Ernst, H. Kahn, A. Heuer, Numerical simulations of carbon and nitrogen composition-depth profiles in nitrocarburized austenitic stainless steels. Metall. Mater. Trans. A **45**(10), 4268–4279 (2014)
84. E. Vyazmina, J. Sheng, S. Jallais, L. Bustamante-Valencia, P. Bruchet, F.-P. Richard, Carbonitriding: Kinetic modeling of ammonia and acetylene decomposition at high temperature and low pressure, in *Proceeding of 24th IFHTSE Congress/European Conference on Heat Treatment, June 26–27, 2017*, Nice (2017)
85. M. Krasowski, *Nitrogen assisted low-pressure carburizing of steel*. PhD thesis, Lodz University of Technology, Lodz (2011). (in Polish)
86. P. Kula, R. Pietrasik, K. Dybowski, S. Pawęta, E. Wołowiec, Properties of surface layers processed by a new, high-temperature vacuum carburizing technology with prenitriding - PreNitLPC. Adv. Mater. Res. **452–453**, 401–406 (2012)
87. S. Pawęta, *Technical and economic models for the implementation of the effective PreNitLPC vacuum carburizing process*. PhD thesis, Lodz University of Technology, Lodz (2014). (in Poland)
88. M. Yan, Study on absorption and transport of carbon in steel during gas carburizing with rare-earth addition. Mater. Chem. Phys. **70**, 242–244 (2001)
89. R. Yin, Carburization performance of Incoloy 800HT in CH4/H2 gas mixtures. Mater. Sci. Eng. A **380**(1–2), 281–289 (2004)
90. N. Ryzhov, A. Smirnov, R.S. Fakhurtdinov, L. Mulyakaev, V. Gromov, Special features of vacuum carburizing of heat-resistant steel in acetylene. Metal Sci. Heat Treat. **46**(5), 230–235 (2004)
91. N. Ryzhov, A. Smirnov, R. Fakhurtdinov, A. Mordovin, Vacuum carburizing of chromium-nickel steel. Metal Sci. Heat Treat. **50**(1–2), 63–67 (2008)

92. R. Gorockiewicz, A. Adamek, M. Korecki, The LPC process for high-alloy steels. Gear Solutions **8**, 40–51 (2008)
93. N. Ryzhov, R. Fakhurtdinov, A. Smirnov, L. Fomina, Analysis of methods of carburizing of gears from heat-resistant steels. Metal Sci. Heat Treat. **52**(5), 260–265 (2010)
94. A. Sugianto, M. Narazaki, M. Kogawara, A. Shirayori, S. Kim, S. Kubota, Numerical simulation and experimental verification of carburizing-quenching process of SCr420H steel helical gear. J. Mater. Process. Technol. **209**(7), 3597–3609 (2009)
95. E. Nart, K. Genel, Investigation of carburising treatment of injector flange in diesel engine. Mater. Design **31**(4), 2110–2117 (2010)
96. D. Kim, H. Cho, W. Lee, K. Cho, Y. Cho, S. Kim, H. Han, A finite element simulation for carburizing heat treatment of automotive gear ring incorporating transformation plasticity. Mater. Design **99**, 243–253 (2016)
97. M. Korecki, J. Olejnik, P. Kula, R. Pietrasik, E. Wołowiec, Multi-purpose LPC+LPN+HPGQ 25 bar N2/He single chamber vacuum furnaces, in *Proceedings of 26th ASM Heat Treating Society Conference (HeatTreat 2011), Oct 31–Nov 2, 2011*, pp. 309–314, Cincinnati (2011)
98. M. Korecki, E. Wołowiec-Korecka, W. Stachurski, A. Brewka-Stanulewicz, M. Bazel, M. Sut, Efficiency and productivity. Successes of vacuum carburizing applications based on CaseMaster Evolution and UniCase Master furnaces, in *Proceedings of 19th Seco/Warwick Seminar: Modern Trends in Heat Treatment, Sept 13–15, 2016*, pp. 145–153, Mierzęcin (2016). (in Polish)
99. M. Korecki, E. Wołowiec-Korecka, D. Glenn, Single-piece, high-volume, low-distortion case hardening of gears, in *Proceeding of AGMA Fall Technical Meetng, Oct 18–20 (2015)*, pp. 1–9, Detroit (2015)
100. M. Korecki, A. Brewka, The method of vacuum injection carburizing (LPC) of elements made of iron alloys and other metals. Patent PL 240533, Seco/Warwick, Poland (2022)
101. M. Korecki, W. Fujak, J. Olejnik, M. Stankiewicz, E. Wołowiec-Korecka, Multi-chamber furnace from vacuum carburizing and quenching of gears, shafts, rings and similar workpieces. Patent US 9,989,311, Seco/Warwick, USA (2018)
102. A. Brewka-Stanulewicz, *Impulse low-pressure carburizing for high efficiency production with continuous single-piece flow of the parts*. PhD thesis, Lodz University of Technology, Lodz (2022). (in Polish)
103. D. Herring, Gas nitriding: something old and something new. Ind. Heat. **85**(8), 10–11 (2007)
104. P. Kula, R. Atraszkiewicz, E. Wołowiec, Modern gas quenching chambers supported by SimVaC Plus Hardness application. Ind. Heat. **3**, 55–58 (2008)
105. E. Wołowiec, *Computer Design of Heat Treatment Processes* (Lodz University of Technology, Lodz, 2013). (in Polish)

Chapter 4
Development of Quenching Towards Quality Improvement

Nothing is worth doing unless the consequences are serious.

George B. Shaw, 1856–1950

4.1 Quenching Deformations as a Quality Determinant

There is a group of consumers for whom product price is of secondary importance, preferring quality in its broadest sense. Although being a minority group in the market, it is significant enough that entire product lines (cars, machines, tools) are being developed focusing on this aspect of products. Just as numerous as the consumer group of luxury goods are industries where safety is an absolute priority (aviation, medicine). The aspect of quality and reliability always comes to the fore there, overshadowing economic aspects and impinging on consumer expectations, also in relation to thermo-chemical treatment. For the automotive and aerospace sector, these are primarily the expectation of manufacturing accuracy and maintaining repeatability of mass-produced parts, issues already addressed in the previous chapter. Chapter 3 discusses this issue in terms of carburising, that is, the first stage of surface hardening of equipment and machine parts. In the case of quenching, this is an even broader issue and primarily covers the aspects of quenching deformations and how to reduce them.

PN EN ISO 4885:2018-05 standard defines quenching as a *the process of hardening of an object, obtained after austenitisation, by cooling under such conditions, in which the transformation of austenite to martensite or bainite takes place* [1]. The use of an appropriate quenching temperature ensures that fine-grained austenite is produced and, fine-grained martensite on cooling. Depending on the type of steel and the method of cooling, a distinction is made between direct-quench hardening, (single and double) step quenching and quenching with isothermal transformation.

4.1.1 Direct-Quench Hardening

Direct-quench hardening is a thermal procedure consisting of heating the steel to a temperature of 30–50 °C higher than Ac_3, annealing the workpieces therein for the time necessary for the austenite transformation to take place, and then cooling in the quenching medium (water, oil, gas). Whether the expected structure after hardening is achieved is preconditioned by the cooling rate; for martensitic quenching, this must be greater than the critical rate for the given material, so that no other phase transformations have time to take place by the time the martensite transformation temperature is reached, and so that the overcooled austenite can only transit to martensite at a temperature below the temperature of martensite transformation M_f. A number of process parameters need to be taken into account to perform quenching correctly, as each of them (type and parameters of the cooling medium, surface area, shape and mass of the cooled parts) has a significant impact on the end result. Normal quenching can be applied to a variety of parts made of alloy steels and to uncomplicated parts made of carbon steels, as parts with complex geometries hardened by this method easily break due to hardening stresses.

4.1.2 Step Quenching

Step quenching is used when machining the parts with complicated shapes; cooling is performed at two rates here. The advantage of this quenching is that the rate of transformation of austenite to martensite is considerably reduced, resulting in lower quenching stresses. Once the quenched part has been annealed at the appropriate temperature, it is first cooled at a high rate in a medium having a temperature higher than the temperature of martensite transformation, it is held at this temperature for some time in order to equalise the temperature across the section, and then cooled in air or oil. This type of quenching is recommended for fine parts, as larger objects, due to the large amount of heat accumulated in the material, may not be quenched to martensite, but to transitional structures.

4.1.3 Quenching with Isothermal Transformation

The first isothermal quenching processes differed from direct-quench hardening in that cooling was carried out in salt or metal baths, at a temperature above the martensitic point. After rapidly passing through the temperature range of 450–550 °C, the component was sustained at temperatures of 200–250 °C, where the overcooled austenite transitioned to a bainite structure. Since the transformation of austenite to bainite takes place slowly, quenching with isothermal transformation ensures, amongst others, the almost complete removal of the quenching stresses that cause deformation and fracture of workpieces. Nowadays, quenching with isothermal transformation is often carried out in gas, eliminating the need for special bathing equipment.

4.1.4 Surface Hardening Treatment

Apart from the volumetric quenching methods described above, there are also several surface hardening methods: flame hardening, induction hardening and bath hardening. A common characteristic of all surface hardening methods is that only the surface layer with a preset thickness is heated to a temperature above Ac_3 or Ac_1. The core of the plastic remains unheated and does not change its properties after hardening. The aim of surface hardening is to obtain a hard surface with a soft and resilient core.

Flame hardening consists in heating the surface layer, e.g. with oxy-acetylene burners, followed by rapid cooling under water spray. The thickness of the hardened layer can be adjusted by changing the peripheral speed of the rotating part or by varying the distance of the water spray from the surface.

During induction hardening, the workpiece is placed inside an inductor, which is a coil rolled from a copper tube, along which a high-frequency current flows, creating an alternating magnetic field around it. Thanks to the phenomenon of induction, eddy currents are created on the surface of the workpiece, which, meeting the resistance of the material, heat up the surface of the workpiece very quickly, while the core of the material remains unheated. The depth of the heated layer depends primarily on the frequency of the current, the power of the generator and the heating time. Assuming that the generator power and heating time are constant, the thickness of the heated layer depends on the current frequency, namely the lower the current frequency, the thicker the layer. Flat and cylindrical items are subjected to induction hardening.

Bath hardening consists in immersing the workpiece in a metal (lead, bronze, cast iron) or salt bath, heated to a temperature of at least 50–100 °C higher than the hardening temperature. Metal baths have a high thermal conductivity, so that there is rapid surface heating of the workpiece, while the core remains unheated or heated only to a temperature below the transformation temperature. After immersion in water or oil, a hard surface layer and a ductile core are obtained. Bath hardening can be used for parts hardened completely or only at the ends, such as shaft journals.

4.1.5 Deep Freezing and Tempering

For steels whose end-of-martensitie-transformation temperatures M_f lie below the temperature of 0 °C, sometimes immediately after hardening a sub-zero treatment, known as deep freezing, is applied to bring about a complete martensite transformation and thus reduce the amount of residual austenite. Regardless of whether or not this stage is employed, tempering is applied after hardening, which involves heating the steel to a temperature below Ac_1, holding there and then cooling to ambient temperature. Depending on the temperature used, a distinction is made between [2]:

- low-temperature tempering, carried out in the temperature range of 150–250 °C, to remove hardening stresses while maintaining high hardness and abrasion resistance;
- medium-temperature tempering, carried out in the temperature range of 250–500 °C, to maintain high strength and elasticity, albeit at the expense of reduced hardness;
- high-temperature tempering is carried out in the temperature range of 500 °C–Ac_1, to achieve high impact strength.

A good coolant should dissipate heat quickly at 500–650 °C, but relatively slowly in the range of 200–300 °C. In the range of the martensite transformation (150–300 °C), there is no need for a high cooling rate, as the overcooled austenite can only turn into martensite in this range and a high cooling rate is not needed. Moreover, martensite is forming, the steel experiences a sudden increase in volume (up to 3% greater than that of austenite) and in the event of a sudden transformation, high stresses may develop, leading to geometrical deformations of the components and even hardening cracks.

4.1.6 Residual Stresses

Internal stresses, referred to in the literature as pre-stress or residual stress, are balanced within a defined volume of the component without the contribution of external forces. In 1930, N. Dawidenkow took the size of this volume as the basis for the classification of stresses, distinguishing the following types of stresses [3, 4]:

- type I stresses (macro stresses), balancing in areas commensurate with the dimensions of the part or specimen being examined,
- type II stresses, balancing in areas of a small number of grains or even in the part of a single grain contained between slip planes and amongst blocks of the mosaic structure; changes in the distance of crystallographic planes in very small volumes of material are assumed as the cause of these stresses; type II stresses cause X-ray diffraction at the Bragg angle corresponding to a given distance of the lattice planes,
- type III stresses, balancing in the areas of the individual crystalline lattices (in the submicro-volume); may be the reason for the high hardness and strength of heat-treated alloys. In solid solutions, third-degree stresses arise due to the different diameters of the atoms that make up the crystallographic lattice.

The cause for the appearance of residual stresses in a solid are mechanical or other interactions, after removing which a non-uniform state of deformation remains in the body. In the case of the surface layer, these interactions are closely related to the technology of its formation (crushing, temperature change, structural transformations, changes in chemical composition, phase transformations, diffusion processes, etc.). Hence, it is of considerable importance for the assessment of the strength of the component to determine residual stresses in macro-areas.

4.1.7 Quenching Deformations

The deformations introduced into a component during its shaping and machining lie at the base of its geometric distortions after heat and thermo-chemical treatment. Deformations of the product exceeding the tolerances described in the design drawing need to be removed, e.g. by grinding, or even cause the item to be rejected from further production. Authors of numerous papers on the subject indicate various causes of the stress and deformation phenomena occurring in the material after heat and thermo-chemical treatment. They raise the following issues in particular:

- Uneven cooling rate. The main cause of the volumetric change in the material is the transformation of the austenite phase into martensite and the volume of the martensite cell varies with the carbon content, according to the relationship described by A. Gulyaev [5].[1] As a result of this phenomenon, strong compressive stresses develop in the austenite not transformed during cooling, which inhibit or completely stop further transformation, while the uneven cooling rate of the workpiece causes the martensite transformation to take place at different rates (to different degrees) in the volume of the material [8, 9]. The consequence is uneven volumetric changes in different parts of the component. Also, an excessive increase in cooling intensity leads to an increase in quenching deformations due to generating thermal stresses (the higher the cooling intensity, the greater the temperature gradients along the length or in the cross-section of the workpiece). As a precautionary measure, many researchers suggest isothermal quenching in the temperature range of the occurrence of bainite transformation, which allows to obtain mechanical properties similar to standard quenching and tempering, while reducing the degree of quenching deformations.
- Asymmetry of cross sections of hardened parts (uneven mass distribution). Uneven mass distribution causes the components to cool at different rates, and therefore a non-uniformity of heat uptake occurs, interfering with the expected uniformity of cooling. The proposed remedy for this situation is to change the design of the element or to reduce the asymmetry and to avoid a large variety of cross-sections and one-sided indentations, although it is obvious that such a solution is not always possible [10].
- Lack of transition radii and abrupt changes in cross-sectional dimensions. Intense stresses can occur at points of abrupt changes in the cross-sectional dimensions and are thus particularly prone to the appearance of quenching deformations.

[1] The martensite cell is 3% larger than the austenite cell at a carbon concentration of 0.8% and 4% larger at a carbon concentration of 1% [5–8].

A. Rutkowska proposes—as a precautionary measure—to use gentle radii and transitions in the workpieces, and if the design of the workpiece does not allow for such a solution—to use pre-treatment taking into account the technological nature of the heat treatment [10].

- Insufficient hardenability in relation to the section size and the type of quenching coolant adopted. A. Rutkowska also points to insufficient hardenability in relation to the section size and the adopted quenching coolant as a possible cause of deformation. A proposed remedy is to consider changing the grade of steel or the method of hardening (step quenching, intermittent quenching, isothermal quenching) [10].
- Insufficient homogeneity of the material (segregation, non-metallic inclusions) or surface defects of metallurgical origin (decarburising, forging lap or closure).
- Misalignment of elements. Both, A. Rutkowska and A. Olofsson point out that the positioning of components in the furnace is an important factor for correct hardening and inadequate positioning in the furnace leads to uneven heating, cooling and the formation of component deformations under its own weight. According to A. Olofsson, vertical alignment is generally associated with more deformations, while horizontal alignment is associated with significantly less roundness deformation, although with an increase in flatness errors [11].
- Unsuitable cooling conditions (type of coolant, movement of the medium, temperature, one-sided washing, immersion method). Many researchers point out that an important cause of stresses and strains may be inappropriate cooling conditions, both in terms of the wrong choice of coolant type, quenching temperature and the way in which the cooling medium washes over the component during quenching. An inappropriately chosen hardening parameter determines the inappropriate course of the whole process, in particular the wrongly selected type of cooling medium, and thus the inadequacy of the rate of heat removal from the component can be a major factor in generating deformation.
- Gas flow velocity and pressure in gas quenching. Although gas is a cooling medium with a more uniform heat removal than oil, S. Wang et al. showed that quenching in gas at constant cooling gas pressure and velocity leads to a large temperature difference between the surface and the core, introducing non-uniform thermal stresses and causing deformations [12]. According to the researchers, with the current availability and state of the art of computational techniques, an efficient and cost-effective way to optimise these parameters is to integrate various computational tools, such as computational fluid dynamics (CFD), finite element analysis (FEA) and microstructural modelling, in order to perform numerical tests for a specific type of component (e.g. a type of gear) prior to hardening. This is particularly important in high-volume and mass production, where an error translates into thousands of manufactured parts.

Summarising the existing state of knowledge in the literature, amongst the numerous causes of stresses and strains after heat treatment given by researchers of this phenomenon, the most important of these should be considered:

1. the rate and uniformity of cooling in the material to be cooled,
2. cooling conditions (type of coolant, movement, temperature, one-sided washing, immersionmethod),
3. arrangement in the space of the hardening chamber,
4. workpiece geometry and size.

Quenching deformations disrupt the correct geometry of the component. In the caseof machine parts manufactured with high precision, such as toothed components the correct gear geometry has a key impact on its operating parameters: service efÞciency (energy losses), vibration and noise. Hence, quenching deformations mus be corrected in most cases, e.g. by grinding. Unfortunately, the correction of quenc ing deformations is one of the most costly processes, as it requires the use of dimensional accuracies and the use of expensive tools to work with material with hardness of the order of 64 HRC, with allowances to be removed reaching 0.1 or more. An analysis of the costs of removing quenching deformations carried o in 1995 by the IWT (Institut fur Werkstofftechnic) in Bremen for German industry determined their amount at around approx. e850 million/year in the gear manu- facturing segment for the automotive industry and e1 billion/year for the bearing industry [13]. By transferring the above Þgures to the global industry and taking int accountcontemporary production growth, the current magnitude of these costs can be estimated at around e20 billion. This places a huge burden on the industry, bu the same time it also has the potential for achieving substantial savings.

4.2 Gas and Oil Quenching

In modern industrial practice, quenching is mostly made in oil. Oil is a commo and well-known quenching medium, although its disadvantages are also known, stemming from the three-phase nature of cooling, the different cooling rates in the phases and the lack of any controllability of the process. If it is taken into acco that each of the phases occurs at different times, at different locations in the quenc part, the large and unique deformations associated with oil quenching are obviou A separate economic and ecological issue is the need to wash parts after quenc in dedicated washers, where various chemicals, increasingly difÞcult to accept b environmental regulations, are used [14].

The contemporary literature on oil quenching focuses on the characterisation o new oils, the quenching deformations associated with oil quenching and ways t predict this phenomenon [15–18].

Gas as a cooling medium started to be used along with vacuum heat treatm technology At Þrst, it accelerated the cooling of the charge by natural convectio and thermal conductivity, but with the development of vacuum technology, forced cooling by circulating gas at elevated pressure in a closed system began to be u which multiplied the rate of gas cooling, although it was initially signiÞcantly slower compared to oil quenching. An unquestionable major advantage of gas quenchin over oil quenching is the possibility of reducing quenching deformations, hence much attention has been given to research into increasing the rate of gas cooling (and thus extending the spectrum of its applications) [19]. Today, the basic cooling gas in heat treatment is nitrogen, due to its easy availability, low costs and relative neutrality towards the materials being treated. For some steels, noble gases (argon, helium) are used that are completely neutral to the process, although with different cooling capabilities. In the 1990s, high-pressure gas quenching technology (HPGQ) was so refined that equipment to implement it began to be installed in industry, including the bulk industry [20, 21]. HPGQ markedly improved the quality of the quenching results especially due to deformations. Gas cooling is single-phase and more uniform, plus the rate of cooling can be freely adjusted by changing the gas pressure (density) and inflow rate, making the process controllable. Modern gas quenching systems using nitrogen or helium at a pressure of 25 atmospheres (2.5 MPa) have the efficiency of oil cooling and can fully replace it, taking over the same range of component types and steel grades (cf. Fig. 4.1). Moreover, gas quenching eliminates the need of step washing and is completely environmentally friendly [22].

Irrespective of the type of cooling medium (oil, gas), the intensity and uniformity of cooling of the workpieces laid in the form of a charge during treatment processing remains a separate issue. The quenching intensity determines the core hardness and effective layer thickness, while the uniformity is responsible for the scatter of these parameters in the workpieces within the charge and for the amount of quenching deformations on individual parts. Charge cooling has an important common characteristic irrespective of the device or the cooling medium (liquid, gas), namely that from the perspective of the individual component in the charge, it is directional cooling. This means that the individual cooled component is influenced

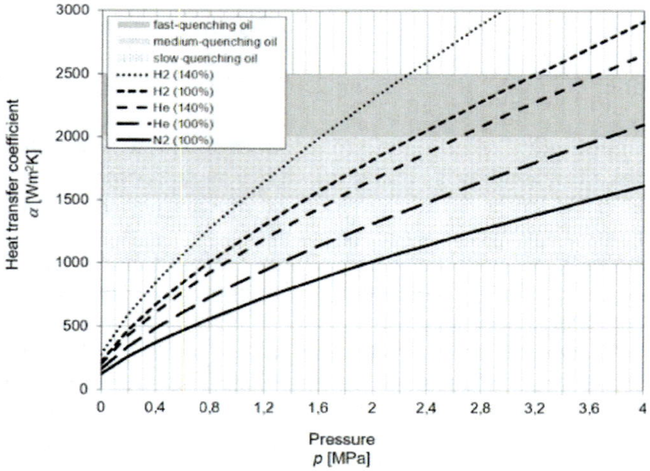

Fig. 4.1 Chart of heat transfer coefficient of cooling gases (hydrogen, helium, nitrogen) depending on pressure and flow rate and in relation to available quenching oils [25]

by the coolant, which flows from one direction—to another (bottom-up, right-left, etc., not necessarily in a straight line) and, as a consequence, is cooled at different rates at different locations. Surfaces on the inflow side are cooled more intensively than those on the outflow (or covered) side, and the difference can even be several times higher [22–24].

As a result of uneven cooling, large temperature differences occur between different locations on the surface or inside the workpiece, resulting in the arising thermal stresses and inhomogeneous microstructure transformations, which ultimately lead to large deformations in the geometry of the hardened workpiece. In addition, a stream of cooling medium in the charge is dispersed and each component is affected by different local cooling conditions. In conclusion, charge hardening (especially in oil) is an uncontrollable and uneven process that results in large deformations within a single workpiece as well as a high degree of uniqueness throughout the charge. The use of gas as a cooling medium can reduce the magnitude of deformations compared to oil, due to the single-phase nature of convective cooling in gas, but all the negative consequences of charge hardening still remain [26]. Furthermore, the charge heat treatment model brings with it additional issues related to quality control, logistics and costs. Monitoring and reporting of the heat treatment process applies at most to the charge (group of parts) and it is not possible to supervise the process from the level of a single part. This makes it much more difficult, if not impossible, to implement more demanding quality assurance systems.

To summarise the traditional *case hardening* technology used on a mass scale, its advantages in terms of powerful performance and minimum costs of the process itself must be appreciated. It is a technology that has been perfected for decades and mastered sufficiently, however, its opportunities for development and improvement have been exhausted. The rising needs and demands of modern industry for improved quality, repeatability, flexibility, speed, process continuity, reduced overall production costs and environmental neutrality could not be met in the traditional way. A new approach was needed to heat treatment technology and equipment that would not only fill the gap that already existed but would be a basis for further development in the future. An approach was needed that opened up new opportunities in three sensitive categories:

1. improved quality and repeatability of results,
2. reduction of quenching deformations,
3. integration of heat treatment equipment in the continuous mass production chain.

4.3 Unit Hardening

In principle, most of the weaknesses and limitations of traditional heat treatment can be linked to its charge nature, so this factor had to be eliminated and replaced by its opposite, the single-piece flow treatment model. Unlike carburising, quenching is a short-lived process, which allows it to be considered on a case-by-case basis.

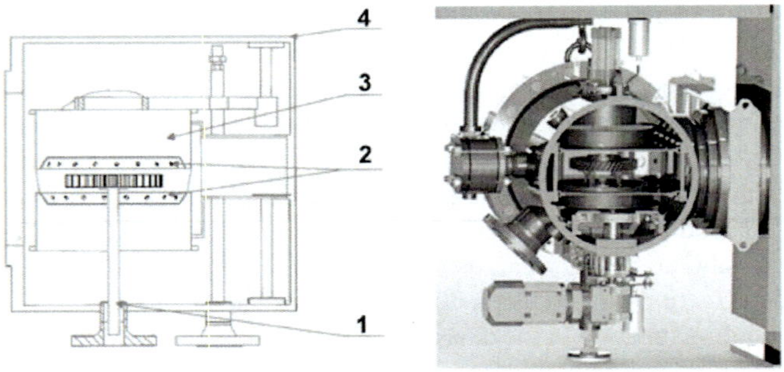

Fig. 4.2 High-pressure gas unit hardening chamber: 1. rotating table, 2. cooling nozzles, 3. cooling manifold, 4. unit hardening chamber. Reproduced with permission of Seco/Warwick

The concept of single-piece flow treatment in heat treatment for mass industry emerged in the scientific and industrial space in 2010, however, until the mid-2010s, no device had been developed that would fulfil this idea in 100% [27, 28]. The first implementation works were recorded in 2015 [14].

The solution proposed by M. Korecki et al. adapts the cooling gas inflow method individually to the geometry of the hardened component [29, 30]. The basis of the system is a system of cooling nozzles surrounding the component and providing an even flow of cooling gas from all sides (3D) and a table that rotates with the component to aid uniformity of cooling (Fig. 4.2). A nozzle cooling system in nitrogen at pressures of up to 1.5 MPa allows to achieve cooling intensities comparable to oil systems, without the need for using helium. The whole process enables the cooling parameters and the configuration of the cooling nozzles to be optimally adapted to the shape of the workpiece in order to homogenise the cooling, which has a direct effect on reducing deformations and achieving highly repeatable results within the batch. Experimental studies have shown that the process of unit gas hardening of a component with the diameter of 200 mm and mass of 2 kg can be carried out in a time not exceeding 60 s. The size and weight of the examined components cover the technical specifications of the gearbox wheels of all passenger cars and most trucks. The length of the hardening process corresponds to the production cycle of CNC machine tools for gear production, where the production cycle lasts approx. 60 s. This makes it possible to handle the stream flow of individual components from the CNC machine tool and thus integrate into the continuous production chain (lean manufacturing) eliminating any logistic operations. By changing the method of high-volume heat treatment from the charge method used until now (multiple workpieces processed in baskets or trays) to single-piece flow, comprehensive and integrated quality control and management at the level of an individual workpiece is possible, which was not possible with charge heat treatment and constituted a gap in the production chain.

Comparative studies of the magnitude of geometric deformations after quenching by different methods show that unit hardening causes the least amount of quenching

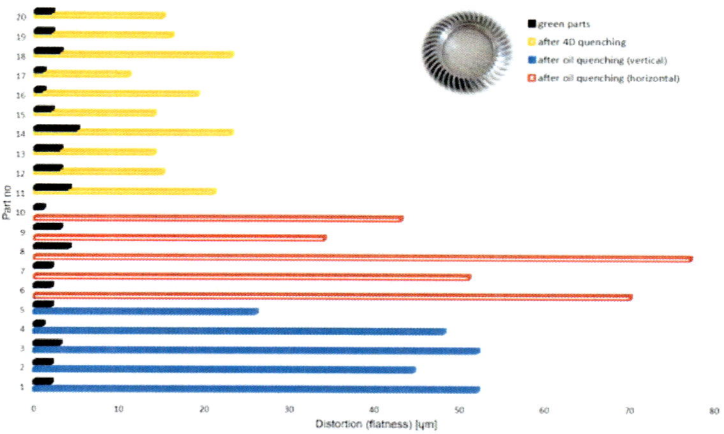

Fig. 4.3 Comparison of geometric deformations (flatness) of components after oil and gas unit quenching

deformations (at the level of the range of capabilities of hardening presses) and in some cases can even reduce the deformations introduced into the component at the shape-forming stage [31]. The Figs. 4.3, 4.4 and 4.5 show three of the more than a dozen investigated geometric deformations of machine parts after thermo-chemical treatment (low-pressure single-piece flow carburising with unit hardening): flatness,[2] roundness[3] and axial runout.[4]

4.4 Further Directions of Development

Although low-pressure single-piece flow furnaces—at the current stage of development—score the lowest in terms of process economics and productivity, they outperform other solutions in the area, especially for accuracy and repeatability of results, integration in the production line and elimination of material logistics, reduction of deformations and deformation removal costs, and control of the process flow and results for each individual part (see Table 3.6). They fit in with the quality requirements of modern industry and, although the direct costs of the process may be

[2] Flatness deviation—the distance between two planes parallel to, tangent to and containing the observed element. (acc. to PN EN ISO 1101:2012 standard).

[3] Roundness deviation—the distance between two concentric circles tangent to the elements observed and containing the element. (acc. to PN EN ISO 1101:2012 standard).

[4] Axial runout deviation—the maximum difference of the distance of the plane measured at the selected radius on the test surface to the plane perpendicular to the base axis. (acc. to PN EN ISO 1101:2012 standard).

Fig. 4.4 Comparison of geometric deformations (axial runout) of components after oil and gas quenching

Fig. 4.5 Comparison of geometric deformations (roundness) of components after oil and gas quenching

higher, they result in a dramatic reduction in production costs beyond heat treatn with Þnishing treatment, parts logistics (no charge tooling, ability to integrate ir continuous production lines) and quality assurance. Thus, ultimately, they also g an advantage in terms of costs.

Nowadays, surface hardening solutions in low-pressure single-piece ßow uni with unit gas quenching are in the area of interest of the automotive industry

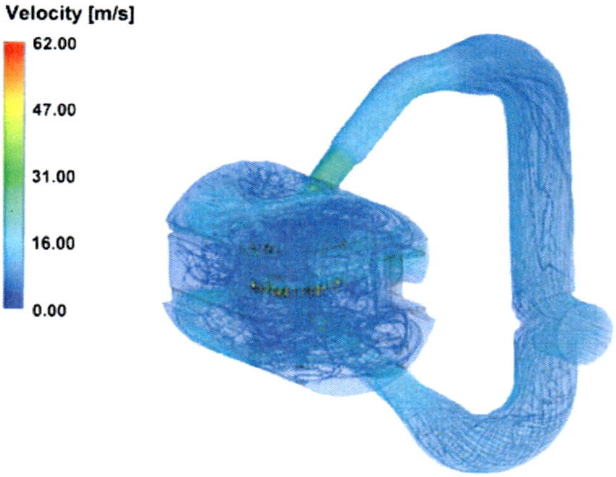

Fig. 4.6 Numerical model of a unit quenching chamber for the study of dispersion of gases and their cooling capacity

aerospace industry. They are an interesting option for *micro-factory*, i.e. local factories, where a wide range of items is produced in the same factory for the local market (local production of cars, buses, machine parts, etc.).

Equipment for implementing the unit quenching method is designed and manufactured in the spirit of the Industry 4.0 concept. It is only possible to design correctly the system of cooling nozzles, tailored to the individual component geometry, by way of modelling and multiple cooling simulations to select the optimum variant. This is confirmed by the works of J. Sawicki et al. [32, 33], in which before a quenching chamber for unit hardening had been made, a numerical model of the quenching chamber was made and cooling conditions in the quenching chamber were analysed, in order to optimise its design (Fig. 4.6). Numerical calculations were carried out to optimise the configuration of the arrangement of cooling nozzles in the quenching chamber for their diameter size, shape, number and method of arrangement. Numerical models of cooling collectors were made. The cooling capacity (values of the average heat transfer coefficient) of a collector with a full arrangement of nozzles, with different active and inactive nozzle configurations and different nozzle distances, was simulated to find the collector with the most effective cooling capacity.

This shows that the further development of quenching methods will be closely linked to the applications of computational techniques, both in the area of designing optimal heat removal from the component, reducing quenching deformations and building equipment to conduct them.

A separate direction will be the development of knowledge on the use of cooling media. It can be anticipated that gas quenching will expand its range of application, with the development of equipment capable of compressing gases to ever higher pressures. The more widespread use of hydrogen as a refrigerant gas is also envisaged. Hydrogen is the gas with the best heat transfer capabilities and the limited number of applications derives from its flammable and explosive nature. Works are however underway over the methods of storing hydrogen and using it safely, for instance, in the energy industry, and the solutions developed will be transferred to other areas of technology. Also, the civilisation's growing concern for the ecological aspects of production may be a reason to move away from traditional oil quenching towards gas quenching, as long as there are ready-made solutions for gas quenching with cooling intensity analogous to oil.

The main flywheel of progress will remain the need to reduce or at least control quenching deformations and thus ensure the quality and reduce unnecessary costs at the stage of production.

References

1. International Organization for Standardization. Ferrous materials—Heat treatments vocabulary. Standard ISO 4885:2018, ISO, EU (2018)
2. F. Staub, J. Adamczyk, Ł. Cieślak, J. Gubała, A. Maciejny, *Metallography* (Śląsk, Katowice, 1978). (in Polish)
3. N. Dawidenkow, Zur Frage der Sprungartigen Deformation. Zeitschrift fur Physik **61**(1–2), 46–53 (1930). (in German)
4. B. Gajducenko (ed.), *Ostatocznyje napriażenija i ustałost provołoki* (Materiały naucznoproizwodstwiennovo seminaria, Odessa-Moscow, 1967). (in Russian)
5. A. Gulyaev, *Heat Treatment of Steel* (Mashgiz, Moscow, 1953). (in Russian)
6. Z. Wendorff, *Metal Science with Heat Treatment* (PWN, Warsaw, 1964). (in Polish)
7. K. Przybyłowicz, *Metallography* (WNT, Warsaw, 2003). (in Polish)
8. L.A. Dobrzański, *Metals and Alloys* (Int'l OCSCO World Press, Gliwice, 2017). (in Polish)
9. L.A. Dobrzański, *Engineering Materials and Material Design. Fundamentals of Materials Science and Metallurgy.* (WNT, Gliwice-Warsaw, 2006). (in Polish)
10. A. Rutkowska, *Manufacturing Techniques. Vol. 2–Selected Issues of Heat Treatment and Thermo-chemical Treatment.* (Cracov University of Technology, Cracow, 1998). (in Polish)
11. A. Olofsson, S. Jonsson, B. Brash, Hardening distortions related to segregations in crown wheels. HTM J. Heat Treat. Mater. **71**(2), 68–74 (2016)
12. S. Wang, J. Wan, D. Zhang, D. Li, C. Zhang, Towards smart factory for Industry 4.0: A self-organized multi-agent system with Big Data based feedback and coordination. Comput. Netw. **101**, 158–168 (2016)
13. Institut fur Werkstofftechnic, *Economic Consequences of Distortion* (Report, Institut fur Werkstofftechnic, Bremen, 1995)
14. M. Korecki, E. Wołowiec-Korecka, D. Glenn, Single-piece, high-volume, low-distortion case hardening of gears, in *Proceeding of AGMA Fall Technical Meetng, October 18-20, 2015* (Detroit, 2015), pp. 1–9
15. K. Ntalianis, A. Croitoru, (eds), *Applied physics, system science and computers II: Proceeding of the 2nd International Conference on Applied Physics, System Science and Computers (APSAC2017), September 27–29, 2017, Dubrovnik.* (Springer International Publishing, Cham, 2019)
16. A. Surajudeen Adekunle, A. Akanni Adeleke, P. Pelumi Ikubanni, K. Adekunle Adebiyi, O. Adekunle Adewuyi, Effectiveness of biodegradable oils as quenching media for commercial aluminium. Materiali in Tehnologije **54**(5), 607–612 (2020)
17. J. Hájek, Z. Dlouha, V. Průcha, Comparison of industrial quenching oils. Metals **11**(2), 250 (2021)
18. J. Liu, J. Li, Y. Tian, Y. Li, Z. Wang, Effects of vacuum oil quenching on distortion of 42CrMo Navy C-ring. Mater. Sci. Technol. 1–8 (2022)
19. P. Kula, R. Atraszkiewicz, E. Wołowiec, Modern gas quenching chambers supported by SimVaC Plus Hardness application. Ind. Heat. **3**, 55–58 (2008)

20. D. Herring, Atmosphere Heat Treatment: Principles, Applications, Equipment, vol. I (BNP Media Group, Birmingham, 2014)
21. D. Herring, Vacuum Heat Treatment: Applications, Equipment, Operation, vol. II (BNP Media Group, Birmingham, 2014)
22. M. Korecki, J. Olejnik, Z. Szczerba, M. Bazel, Single-chamber HPGQ vacuum furnace with quenching efÞciency comparable to oil. Ind. Heat. 9, 73Ð77 (2009)
23. D. Scott MacKenzie, Z. Li, B. Ferguson, Effect of quenchant ßow on the distortion of carburize automotive pinion gears. HTM J. Heat Treat. Mater. 63(1), 15Ð21 (2008)
24. D. Scott MacKenzie, A. Kumar, H. Metwally, S. Paingankar, Z. Li, B. Ferguson, Prediction of distortion of automotive pinion gears during quenching using CFD and FEA. J. ASTM Int. 6(1), 1Ð10 (2009)
25. M. Korecki, Theoretical and Experimental Methods to Support the Design of Universal, Single-chamber Vacuum Furnaces. PhD thesis, (Lodz University of Technology, Lodz, 2008). (in Polish)
26. D. Scott MacKenzie, Heat treatment of gears: Control of residual stress and distortion. Hea Treat. Progr. 7(4), 47Ð56 (2007)
27. V. Heuer, K. Loser, G. Schmitt, K. Ritter, One piece ßow: Integration of case hardening int the manufacturing line, in Profeedings of Conference on Gears 2010 (TUM Garching, 2010)
28. IHI Machinery and Furnace Co. Ltd. In-line heat treatment!? Next-generation heat treatment equipment. IHI Eng. Rev. 44(2), 2Ð5 (2011)
29. M. Korecki, W. Fujak, J. Olejnik, M. Stankiewicz, E. Wo·owiec-Korecka, Device for individual quench hardening of technical equipment components. Patent US 10,072,315 (Seco/Warwick, USA, 2018)
30. M. Korecki, W. Fujak, J. Olejnik, M. Stankiewicz, E. Wo·owiec-Korecka, Device for individual quench hardening of technical equipment components. Patent EP 3006576 (Seco/Warwick, EU, 2020)
31. M. Korecki, E. Wo·owiec-Korecka, M. Sut, A. Brewka, W. Stachurski, P. Zg—rniak, Precision case hardening by low-pressure carburizing (LPC) for high volume production. HTM J. Heat Treat. Mater. 72(3), 175Ð183 (2017)
32. K. Krupanek, A. Staszczyk, J. Sawicki, P. Byczkowska, The impact of nozzle conÞguration or the heat transfer coefÞcient. Arch. Mater. Sci. Eng. 90(1), 16Ð24 (2018)
33. J. Sawicki, K. Krupanek, W. Stachurski, V. Buzalski, Algorithm scheme to simulate the distortions during gas quenching in a single-piece ßow technology. Coatings 10(7), 694 (2020)

Chapter 5
Development of Nitriding to Reduce Consumption of Process Factors

Simplicity and common sense should characterise planning and strategic direction.

Ingvar Kamprad, 1926–2018

5.1 Process Economics

Economic factors are probably the strongest motivator for technology development. Since production costs are influenced (leaving aside the cost of human labour) by run time and the amount of energy consumed, raw materials and consumables used, the need to reduce all these factors was clearly recognised at the dawn of the twenty-first century. It was estimated that it is necessary to reduce the production costs of parts by 75%, including the duration of the heat and thermo-chemical treatment processes themselves by 50% [1]. Nitriding is an example of treatment whose development fits in with these cost-saving trends.

The nitriding process developed, in the early twentieth century, still plays an important role in many industrial applications. Together with a derivative nitrocarburising process, it is often used in the manufacture of machine components and tools where high-load friction nodes and cyclic loads occur. The manufacturers of automotive and aerospace components employ the nitriding of gears, crankshafts and valve parts because the process enriches the surfaces of the components with hard diffusion layers and the enhanced fatigue strength prevents the formation of surface and subsurface cracks [2, 3]. Nitriding has also become an attractive heat treatment option for manufacturers of tool steels and dies, as it increases the hardness of their surfaces without the risk of deformations and improves the tempering resistance of dies at operating temperatures during forging, extrusion and casting of, for example, aluminium. Firearms manufacturers apply nitriding to the barrels and guides of their products because the process reduces the coefficient of friction, heightens wear resistance and fatigue strength and provides some control over corrosion [4, 5].

The varieties of nitriding used in practice include mainly gas nitriding at atmospheric pressure [2, 4, 6–9], glow-discharge nitriding [10–13], laser nitriding [14], low-pressure nitriding [15–23] and salt-bath nitriding [24, 25] and bath nitriding. In the context of research pursued in the twenty-first century, controlled nitriding, nitriding with economical use of ammonia and low-pressure nitriding are particularly targeted at reducing the costs (of consumption of operating gases or time).[1]

5.2 Gas Nitriding

Nitrogen is transferred from the vapour phase to the solid phase in the nitriding process, with the vapour phase being the nitriding atmosphere obtained from partially dissociated ammonia. The solid phase is the steel subject to nitriding. In order to create favourable conditions for the diffusion of nitrogen into the steel, the rate of dissociation of ammonia must be maintained such that the mass of nitrogen obtained from it is slightly greater than the amount of nitrogen adsorbed on the steel surface. The phenomena in the atmosphere leading to the formation of the nitrided layer concern about the composition of the atmosphere and its influence on the nitrogen potential and nitrogen availability [28]. In contrast, the phenomena taking place at the boundary between the atmosphere and the steel surface result from the relationship between the nitrogen potential and the surface concentration of nitrogen in the nitrided layer and the associated flux of nitrogen flow into the steel. However, the phenomena occurring in the steel stem from the relationship between the surface concentration of nitrogen and its distribution in the nitrided layer (Fig. 5.1) [29]. To summarise, for the general case of gas nitriding, the basic parameters characterising the nitriding atmosphere are the nitrogen potential, the rate of ammonia dissociation and nitrogen availability, while the parameters used to regulate them at constant temperature are the chemical composition and the flow rate of the inlet nitriding atmosphere.

Iron nitrides can be produced as a result of the reaction between iron and ammonia: γ' (Fe_4N), ε (Fe_3N), ζ (Fe_2N), whereas the content of nitrogen in the nitrides produced depends on the temperature and the content of ammonia in the gas phase [31, 32]. The phase system of *iron-nitrogen-hydrogen* has been well studied and described in the literature [33–35], as has been the mode of interaction of ammonia, hydrogen and nitrogen on the iron surface [36, 37].[2] Studies into the thermodynamic equilibrium in the *iron-nitrogen-hydrogen* system and into the determination of the limits of the occurrence of the individual phases of iron nitrides depending on the temperature and composition of gaseous ammonia-hydrogen mixtures have been

[1] Part of this chapter has been reproduced with permission from Ref. [26, 27] Copyright (2019) Springer Nature.

[2] The phase diagram for the *iron-nitrogen* system was first presented by K. Jack [38, 39].

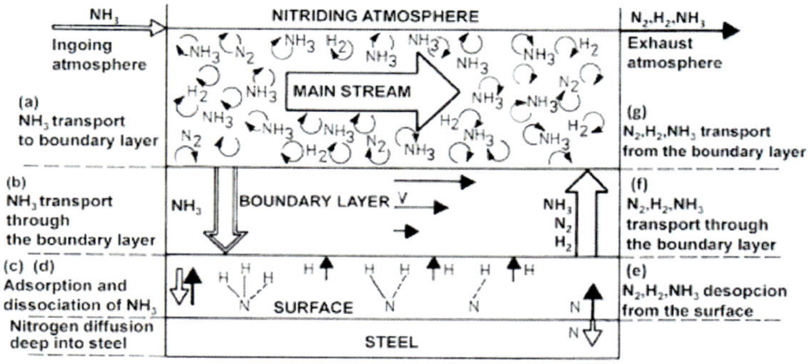

Fig. 5.1 Phenomena that occur during gas nitriding of steel in a nitriding atmosphere: stages **a**, **b**, **f** and **g** supply substrates to form the nitrided zone and remove gaseous reagents resulting from the process; stages **b** and **f**—transport of gaseous reagents through the boundary zone between the main atmospheric stream and the steel surface; stages **c**, **d** and **e** include adsorption, gradual dissociation of the ammonia and desorption of the gaseous reagents. Reprinted with permission from Ref. [30] Copyright (2018) Elsevier

pursued since the 1930s by E. Lehrer [33], and his diagrams were supplemented in the 1990s by L. Małdziński [40, 41]. Between 2001 and 2022, the phase transformations of the *iron-nitrogen* system were subjected to detailed investigations using the latest techniques and research methods of the period [8, 29, 42, 43]. Works of researchers from the Max Planck Institute deserve particular attention in this domain [35, 44–56]. The attempts to optimise the nitriding process reported in this period were almost always undertaken using numerical computational methods such as FEM analysis [57–59].

Gas nitriding has a number of advantages over other thermo-chemical treatment methods, including a low process temperature, the ability to treat a wide range of materials and the ability to control the phase composition and structure of diffusion layers. The correct phase composition of the surface layer ensures the required properties: hardness, wear resistance, corrosion resistance, fatigue strength for the various operating conditions of machine parts or tools. Based on existing models of wear mechanisms, corrosion damage and other surface properties, requirements for the phase composition of the nitrided layer for different operating conditions are formulated. The zone of surface nitrides is essential for creating the properly working friction pairs; it also improves the corrosion resistance of steel. For components operating under high-pressure wear conditions, a multilayer structure—consisting of a zone of surface nitrides with an internal alloy zone of nitriding—is preferred. A continuous zone of internal nitriding is helpful under conditions of high dynamic loading combined with wear [60].

The process of gas nitriding—although complex from the physical and chemical model's standpoint—from the standpoint of industrial implementation is seen as the simplest amongst the surface hardening techniques. A key aspect of the nitriding process is that it does not require a phase change from ferrite to austenite. The lattice structure of ferrite (regular face-centred lattice, bcc, A2) does not change its configuration nor develops into the regular wall-centred lattice (fcc, A1) characteristic for austenite. Furthermore, no subsequent transformation of austenite to martensite occurs. Again, there is no change in the structure of lattice cells and, more importantly, no dimensional change in the elements, only a slight increase caused by the change in the volume of the steel surface due to diffusion of nitrogen. What can cause distortions are the induced surface stresses released by the heat of the process, causing movements such as torsion and bending [3, 61–63].

5.3 Controlled Gas Nitriding Under Atmospheric Pressure

In the 1970s, J. Tacikowski and J. Zyśk observed that failure to ensure the proper availability of nitrogen in a gaseous atmosphere (its excessive amount) during nitriding results in the formation of a brittle ε layer, which has to be removed mechanically, with the cost of removal being high. Since the diffusion processes in a solid can only be controlled by controlling the flux of nitrogen reaching the gas-solid boundary, this control was initially achieved by adjusting the rate of ammonia dissociation: a mixture of NH_3/NH_{3diss} gases was supplied into the furnace chamber with a proportion appropriate to achieve the desired effect. The method proposed by J. Tacikowski and J. Zyśk in the mid-1975s extended the range of the control: other gases, at least one, were added continuously or periodically to the ammonia supplied into the nitriding chamber, such as argon, hydrogen, nitrogen, dissociated ammonia, endo gas, exo gas, decomposition products of liquid organic compounds, water vapour, oxygen or hydrogen chloride [64]. A similar concept of dilution with ammonia gases was postulated in 1978 by Y. Lakhtin and Y. Kogan, drawing attention to mixtures of gases: a mixture of ammonia with argon or nitrogen, mixture of ammonia with its dissociation products or hydrogen, mixture of ammonia with gases containing carbon or oxygen [65].

In a process of controlled gas nitriding where a nitriding atmosphere is obtained from a two-component mixture of ammonia and dissociated ammonia (NH_3/NH_{3diss}), layers of iron nitrides γ' can be obtained for a wide range of ammonia dissociation rates. In a process of controlled nitriding in a nitriding atmosphere obtained from a two-component mixture of ammonia and nitrogen (NH_3/N_2) at high dilutions with nitrogen and in processes at reduced pressure, the phase stability area of the γ' is much narrower, so it is more difficult to produce a monophase layer of iron nitrides γ'. An increase in the dilution rate of the inlet atmosphere with nitrogen is accompanied by a reduction in the phase stability area of γ' layer and an increase in the phase stability area of the ε nitrides. The nitriding method described above has found widespread use in industry and is subject to further development works (*Nitreg* technology) [66, 67].

In 2010, J. Michalski demonstrated that the use of multicomponent nitriding atmospheres in controlled gas nitriding results in measurable economic savings during the implementation of this process [68]. He also proved that nitrogen availability can be a parameter of the nitrogen atmosphere equivalent to the nitrogen potential [28, 67, 69].

Currently, controlled gas atmospheric nitriding can be carried out using four types of inlet atmospheres: a one-component ammonia atmosphere (NH_3), a two-component ammonia atmosphere with dissociated ammonia (aNH_3/cNH_{3diss}), a two-component ammonia atmosphere with nitrogen (aNH_3/bN_2), and a three-component ammonia atmosphere with dissociated ammonia and nitrogen ($aNH_3/bN_2/cNH_{3diss}$), where the coefficients a, b, c denote the percentages of the individual gases [68].

Mathematical models of the change in the composition of the inlet atmosphere for the assumed changes in nitrogen potential were developed by J. Michalski and T. Frączek in 2020 [70].

5.4 Controlled Gas Nitriding with Economical Use of Ammonia

Gas nitriding with the economical use of process gas is a variant of atmospheric nitriding, strongly focused on the economic and environmental aspects of its application. The essence of the method is that the process of controlled nitriding is performed with ammonia alone, and the control of chemical composition of the atmosphere in the furnace retort (and consequently the control of the kinetics of the growth and construction of the nitrided layer) is performed by periodically closing and opening its supply to the retort.[3] The amount of ammonia entering the furnace is controlled by means of a gas analyser, which opens or closes the gas inlet valve at the frequency required to maintain the desired gas composition (desired potential) in the retort.[4] The process was given its name, *ZeroFlow*, from the way it is controlled, and was quickly introduced into industrial practice. Although the literature does not provide a thorough phenomenological description of this process, there are many reports of its practical applications [41, 71–76].

[3] The control of the chemical composition of the atmosphere in the retort, and thus the control of the potential, takes place by changing the rate of ammonia supply to the retort, whereby at certain stages of the process the ammonia supply to the furnace can be reduced to zero.

[4] The concept presented here is not an entirely new solution, as already in previously known methods the atmospheric composition was controlled by changing the ammonia flow rate into the furnace, albeit within a relatively narrow range, without closing the inflow completely (e.g. in some aerospace applications, significant restrictions in the inflow of ammonia were applied with a high drop in potential).

5.5 Low-Pressure Nitriding

The need to reduce the supply of nitrogen during nitriding [15, 77] was underpinning the development of low-pressure nitriding (LPN) in the late 1970s. Research works of this period showed that, at pressures in the range of 400–500 hPa, the structure of the nitrided layer is similar to that observed when nitriding at atmospheric pressure, while the growth of the nitride zone at this pressure is rising markedly. At pressures in the range of 200–270 hPa, the formation of the ε phase is suppressed and the nitride zone is mainly composed of the γ' phase. At pressures in the range of 133–200 hPa, the nitride zone does not form on the surface, and the diffusion layer is only a solid solution of nitrogen α, the thickness of which is 25–30% greater than after nitriding with normal technology at atmospheric pressure (1013 hPa) [16]. The control of the amount of nitrogen supplied to the surface of the workpiece during the process was the essence of the low-pressure nitriding process in an atmosphere of dissociated ammonia, known to have been implemented in Poland in the 1980s and 1990s, in the pressure range of 3–1050 hPa (*Nitrovac'79* technology) [78].[5] The low partial pressure of hydrogen in the process had the effect of lowering the interaction of this element with the surface of the charge, and the layers obtained by this method were more ductile compared to the layers obtained by other technologies [77, 79, 80].

A new way of controlling nitrogen supply in low-pressure nitriding, by segmentation, was developed in 2010. The idea of multi-segment low-pressure nitriding consists in running the process with cyclic dosing of ammonia into the working chamber (saturation stage) and cyclic removal of ammonia (annealing stage) [81, 82]. The stages of saturation are conducted with a continuous flow of ammonia under a constant total pressure in the range of 20–30 hPa, which guarantees a constant and repeatable content of nitrogen in the layer of iron nitrides being produced. Withstanding stages are performed in the absence of ammonia supply to the surface, resulting in an increase in the thickness of the internal nitriding zone (solution zone) with a reduction in the thickness of the layer of iron nitrides. This enables to model the growth of the nitrided layer, in particular to obtain layers with a significantly reduced layer of nitrides γ' and ε. The nitriding processes at reduced pressure, developed by this method, have found their application in the treatment of moulds, dies and other components made from tool steels (*Fine*LPN technology) [23]. The relatively short nitriding processes for these steels, lasting a few hours, produce a thin layer with high hardness, which improves the performance of these tools. In addition, the vacuum-nitrided layers are less brittle than gas-nitrided layers, both in the confined zone of nitride compounds and in the zone of internal nitriding.

[5] The process was executed by placing the elements in a retort of a pit-type furnace containing ammonia, into which sulphur vapours were introduced for a period of 5 sec–60 min., after which the workpieces were subjected to two-stage nitriding in an atmosphere of ammonia or ammonia and sulphur vapours, possibly also the then known gases used in nitriding processes. Nitriding in the first stage was carried out at a pressure at least 10 times lower and for a time at least 5 times longer than in the second stage of nitriding, in which the pressure of the treatment atmosphere was maintained between 3 and 1050 hPa, and the duration of this stage was between 5–30 min. [78].

As nitriding processes are characterised by a stage of incubation, the nitride surface must be activated prior to the supply of ammonia to execute the proc effectively. The activation of the charge surface prior to nitriding is known primarily for parts made of acid-resistant, heat-resistant and stainless steels. These steels particularly prone to the formation of dense, persistent oxides on their surface a failure to remove the oxides prior to the nitriding process makes the formation of nitrided layer on these steels virtually impossible. For carbon and low-alloy steels surface activation is somewhat less important, as these steels are not as sensitiv the formation of oxides in the nitriding process, but the nitrogen saturation process are significantly slowed down by omitting the surface activation. The importance o this stage has been described in numerous papers [81, 83], it is essential to ensure the repeatability of the layers produced. By activating the surface before the nitriding process, the layer structure is unified and the depth of penetration of nitrogen atc is increased by removing oxides and impurities. It is important to activate the surfa of structural steel mainly for short-term nitriding processes (1–3 h). If the stee surface is not activated prior to the nitriding process, this is clearly evident in t characteristics of the nitrided layer, for both low- and medium-alloy steels. For shor process times (of the order of 2 h, at a temperature of 5[illegible]0 °C), the maximum hardness of the layer can be 50% lower compared to the layer on the steel with an activ surface. Differences in effective thicknesses of layers can be as high as 30%.

In the early 2000s, low-pressure nitriding finally found its application areas in th global thermo-chemical treatment industry. Ways to control the supply of nitrogen to the surface in the nitriding process were being developed in the scientific area [30, 81, 84–86]. Similar analyses were pursued in the industrial area for specific stee grades [17, 18, 20, 87–89].

5.6 Unification of the Nitriding Model

Work was undertaken in 2018 to unify the mathematical notation of the kineti equations of nitrogen layer growth for the entire spectrum of pressures applicable practice [26, 30, 86]. It has been pointed out already earlier that in both, low-pressure nitriding processes using pure ammonia and atmospheric nitriding processes in mixtures of aNH_3/bN_2, the analytical models using the nitrogen potential and the corresponding equilibrium surface pressure of nitrogen as an input parameter are n applicable [81]. Therefore, another parameter was sought to relate the parameter o atmosphere to the kinetic effect of the process.

Phenomena in the gas and solid phases

In the general case of gas nitriding, the most relevant parameters influencing t growth kinetics of the nitrided layer and its phase composition are temperature nitrogen potential, the composition and flow rate of the nitriding atmosphere, th chemical composition of the nitrided steel and the condition of surface preparation, stated in the works of H. Larsson, J. Agren et al. [90–93]. For low-pressure nitriding, nitriding is carried out technically at a constant pressure and for a constant-flow rate of ammonia, which means a constant value of nitrogen potential and a constant flux of nitrogen from the nitriding atmosphere to the nitrided surface. A flux of nitrogen can also be controlled by segmenting the process, i.e. splitting it into stages (segments): saturation under conditions of continuous flow of ammonia and subsequent annealing with closed supply of ammonia and constant flow of inert gas at the same flow rate. Alternatively, a flux of nitrogen can be reduced by lowering the nitrogen potential of the nitriding atmosphere or by diluting the inlet atmosphere with an inert gas; in both cases, there will be a change in the rate of ammonia dissociation and consequently in the flux of nitrogen to the nitrided surface [28, 68].

Figure 5.2 shows the rate of nitrogen production from the ammonia dissociation reaction and a variation of ammonia content in the outlet atmosphere as a function of ammonia flow rate. For very low flow rates (range I), the time the ammonia molecule stays on the steel surface is long enough for it to dissociate, and the flow rates of the atmosphere in the main flux and in the boundary zone are equal to each other [94]. The growing flow rate of the atmosphere is accompanied by a rising amount of ammonia flowing into the steel surface. There is still small but increasing number of its molecules in equilibrium with the surface. In this range of flow rates, the rate of nitrogen production is proportional to the flow rate of the inlet atmosphere; the outlet atmosphere contains practically no ammonia, only the products of its dissociation (range I). As the flow rate of the nitriding atmosphere is growing further, a moment comes (point 1) when the time the ammonia molecules are staying on the surface is too short for all the molecules to dissociate. Hence, some molecules do not participate in the dissociation reaction. As a result, the outlet atmosphere contains

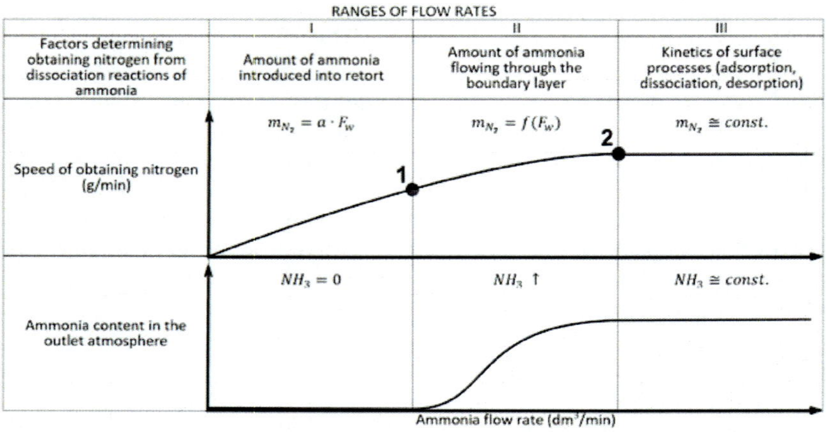

Fig. 5.2 The rate of nitrogen production by dissociation of ammonia and a change in the ammonia content in the outflowing atmosphere depending on the flow rate of the single-component inflow ammonia atmosphere. Adapted with permission from Ref. [30] Copyright (2018) Elsevier

also ammonia, in addition to the products of the dissociation reaction, nitrogen and hydrogen. From this moment (range II), the rate of nitrogen production is determined by the inflow of ammonia through the boundary zone to the steel surface. The rate of nitrogen production then grows monotonically as the flow rate of the inlet atmosphere (of ammonia) is rising, and at point 2 the maximum nitrogen production rate is established. A further growth in the flow rate of the inlet atmosphere (range III) no longer results in an increase in the nitrogen production rate, as manifested in a constant content of ammonia in the outlet atmosphere. In this respect, the rate of nitrogen production is determined by the kinetics of surface reactions. Nitriding processes in conditions of atmospheric pressure are most often conducted at flow rates corresponding to range II, as confirmed by the work of J. Michalski [95]. Low-pressure nitriding processes are performed at flow rates corresponding to range III, as confirmed by the studies of E. Wołowiec-Korecka [96]: under these conditions, the highest nitrogen availability of the nitriding atmosphere is obtained. On the other hand, relatively high surface concentrations of nitrogen are achieved at reduced pressure, resulting in the formation of a nitride layer on the surface of the steels studied, composed of a mixture of $\gamma' + \varepsilon$ phases with a predominance of γ' phases [30].

In the solid phase, the basic phenomena taking place in steel result from the relationship between the surface concentration of nitrogen and the concentration gradient in the nitrided layer [97]. The behaviour of ammonia on surfaces of ferritic and austenitic iron has been extensively described by H. Grabke, and his equations are used to date in many papers [98–100]. The process of constituting the nitrided layer can be divided into two main stages, the first being the saturation of the steel with nitrogen and the second being the formation of a near-surface layer of iron nitrides, with further growth of the solution layer. The phenomena taking place in steel and creating the nitrided layer are preceded by nitrogen adsorption processes. The adsorbed nitrogen *in statu nascendi* can diffuse deep into the substrate or react with iron to form iron nitride. While diffusion is a process progressing in the solid phase, nitride synthesis takes place with the presence of nitrogen from the gas phase. Processes with the participation of the gas phase have lower activation energies compared to processes progressing in the solid phase and are therefore privileged. Reduced pressure further promotes an increase in the rate of ammonia dissociation. This means that in low-pressure nitriding processes there is a tendency to reduce the growth of the solution layer with the growth of the layer of iron nitrides [30].

Nitrogen potential

The notion of nitrogen potential (N_P) as a physiochemical parameter determining the gas nitriding process was introduced by B. Lightfoot and D. Jack in 1973 [101]. Nitrogen potential is not a parameter determining the kinetics of the process, but it serves as a starting point for selecting the optimum variants of nitriding atmospheres in terms of their type and composition (it characterises the state of concentration equilibrium of nitrogen between the nitriding atmosphere and the steel surface). Its value is determined on the basis of equilibrium partial pressures of ammonia and hydrogen in the nitriding atmosphere (Eq. 5.1) and, assuming that the components of the nitriding atmosphere are a mixture of perfect gases, the potential can also be determined from the volume content of the components of the nitriding atmosphere according to Dalton's law (Eq. 5.2).

$$N_P = \frac{p_{NH_3}}{(p_{H_2})^{1.5}} \tag{5.1}$$

where p_{NH_3}, p_{H_2}—partial pressures of ammonia and hydrogen in the outlet nitriding atmosphere.

$$N_P = \frac{V_{NH_3}}{(V_{H_2})^{1.5}} \tag{5.2}$$

where V_{NH_3}, V_{H_2}—volumetric proportions of ammonia and hydrogen in the outlet atmosphere, respectively.

The general form of the equation describing the potential of the nitriding atmosphere obtained from a two-component nitriding atmosphere $a\mathrm{NH_3}/b\mathrm{N_2}$, as a function of the volumetric proportions of ammonia and nitrogen in the inlet atmosphere and pressure, is as follows:

$$N_P = \frac{1 - b - \frac{2}{3} V_{H_2}(2 - b)}{(V_{H_2})^{1.5} p^{0.5}} \tag{5.3}$$

where b—nitrogen content in the inlet atmosphere, p—pressure of the inlet atmosphere (atm.), V_{H_2}—hydrogen content in the outlet atmosphere.[6]

For values of parameters $b = 0$ and $p = 1$, the nitrogen potential Eq. (5.3) takes the form for a single-component inlet atmosphere (100% NH_3) for conditions of atmospheric pressure (5.4):

$$N_P = \frac{1 - \frac{4}{3} V_{H_2}}{(V_{H_2})^{1.5}} \tag{5.4}$$

For a two-component atmosphere $a\mathrm{NH_3}/b\mathrm{N_2}$ as a function of the hydrogen content in the nitriding atmosphere in the conditions of atmospheric pressure, the equation has the following form (Eq. 5.5):

$$N_P = \frac{1 - b - \frac{2}{3} V_{H_2}(2 - b)}{(V_{H_2})^{1.5}} \tag{5.5}$$

In addition, the relationships (5.6) and (5.7) permit to calculate the nitrogen potential of the nitriding atmosphere obtained from a single-component inlet atmosphere (100% NH_3) and a two-component atmosphere of ammonia with nitrogen

[6] Exact conversions are given in the work of J. Michalski, E. Wołowiec-Korecka. *A study of parameters of nitriding processes*. Metal Science and Heat Treatment 61 (2019) [26, 27].

(aNH_3/bN_2), respectively, as a function of the hydrogen content in the nitriding atmosphere below atmospheric pressure:

$$N_P = \frac{1 - \frac{4}{3}V_{H_2}}{(V_{H_2})^{1.5}(p_{NH_3})^{0.5}} \tag{5.6}$$

$$N_P = \frac{1 - b - \frac{2}{3}V_{H_2}(2 - b)}{(V_{H_2})^{1.5}(p_{aNH_3/bN_2})^{0.5}} \tag{5.7}$$

where p_{NH_3}—pressure of the one-component inlet atmosphere of ammonia, p_{aNH_3/bN_2}—pressure of the two-component atmosphere of ammonia with nitrogen.

Rate of dissociation

The control of any gas nitriding process consists in regulating the amount of nitrogen flux generated by the thermal decomposition of ammonia. A measure of this decomposition can be the nitrogen potential, which determines the potential of the nitriding atmosphere from the point of view of the formation of nitrogen phases α, γ', ε according to E. Lehrer's system [33]. However, since in the process of atmospheric nitriding only about 0.64% of the mass of nitrogen formed by dissociation is used to create the nitrided layer, the ammonia atmospheres are used in these processes diluted with an inert gas, e.g. nitrogen or ammonia decomposition products, i.e. dissociated ammonia. For atmospheres with a low content of active nitrogen, i.e. mixtures of NH_3/N_2 and for processes at reduced pressure, the value of the nitrogen potential loses its usefulness in characterising the process capability and the rate of ammonia dissociation becomes important, which is a quantitative parameter determining how much ammonia must decompose to supply the amount of nitrogen *in statu nascendi* necessary to create a nitrided layer and reach equilibrium, commonly termed as nitrogen potential. The rate of dissociation is important for the stability of the nitride phases produced during the nitriding process, even at reduced pressure (Fig. 5.3).

The rate of ammonia dissociation (ω) in a nitriding atmosphere is expressed by the ratio:

$$\omega = \frac{2s}{1 + s} \tag{5.8}$$

where s—the portion of ammonia in the inlet atmosphere that dissociated during the nitriding process:

$$s = \frac{V_{H_2}}{1.5 - V_{H_2}} \tag{5.9}$$

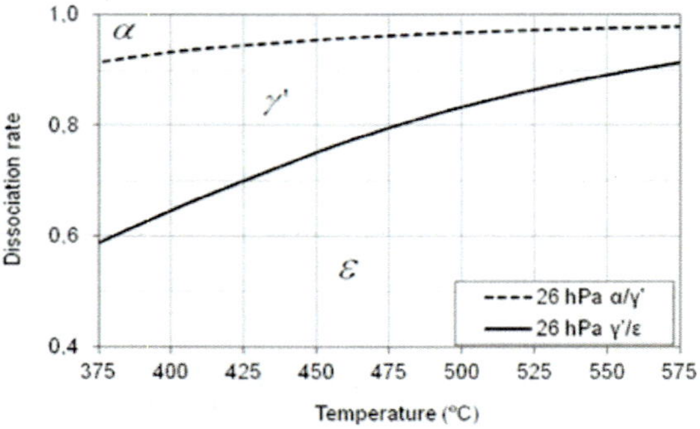

Fig. 5.3 The equilibrium Fe-N system according to E. Lehrer [33], considering the occurrence of the individual phases at reduced pressure (26 hPa): the ε phase is a solid phase over the largest range of ammonia dissociation rates, while the α phase—in the smallest range; the diagrams refer to thermodynamic conditions which are not always the same as kinetic conditions. Reprinted with permission from Ref. [30] Copyright (2018) Elsevier

A formula for calculating the rate of dissociation of ammonia as a function of the hydrogen content of the outlet atmosphere is derived from substituting the relationship (5.8) into Eq. (5.9):

$$\omega = \frac{4}{3} V_{H_2} \tag{5.10}$$

The content of hydrogen, as a function of the rate of dissociation, also follows from the transformation (5.10):

$$V_{H_2} = \frac{3}{4} \omega \tag{5.11}$$

Nitrogen availability

Two basic chemical reactions occur in the nitriding process [98]:

$$NH_3 \rightarrow \frac{1}{2} N_2 + \frac{3}{2} H_2 \tag{5.12}$$

$$NH_3 \rightarrow N_{(Fe_\gamma)} + \frac{3}{2} H_2 \tag{5.13}$$

where $N_{(Fe_\gamma)}$—nitrogen in the solid phase.

The inlet atmosphere of ammonia with a volume of V_{in} is the sum of the volume of ammonia that will dissociate under process conditions and the volume of ammonia that will not dissociate. When the inlet atmosphere is introduced into the furnace chamber, at the process temperature, the volume $V_{NH_{3diss}}$ dissociates according to the

Eq. (5.12), as a result of which its volume doubles. This means that the nitridin atmosphere has a larger volume than the inlet atmosphere and the growth in volu is greater the greater the proportion of ammonia dissociates. The outlet atmosphe outside the furnace chamber is subject to no changes. Thus, in terms of composit it is the same as the nitriding atmosphere and the volume of the outlet atmosphe equal to the volume of the nitriding atmosphere [58]. The outlet atmosphere is a mixture of ammonia, hydrogen and nitrogen in proportions resulting from the chemical reaction of dissociation, hence by measuring the content of one of the components the outlet atmosphere, i.e. ammonia, hydrogen or nitrogen, the nitriding atmospher can be described quantitatively.

In a closed system, with no flow, the reaction of ammonia dissociation (Eq. 5.12) proceeds virtually until it is completely decomposed and an atmosphere is produced consisting of the products of the nitrogen and hydrogen dissociation reaction. Onl very low concentrations of nitrogen in iron can be achieved under such conditions atmospheric pressure. Atmospheric gas nitriding is hence executed in an open syst with the flow of atmosphere, whereas the chemical composition of the nitriding atmosphere is being controlled by varying the flow rate of the inlet atmosphere. compensates for the loss of ammonia caused by the reactions (5.12) and (5.13). Thus, by changing the flow rate of the inlet atmosphere, it is possible to regulate the r of ammonia dissociation and, as a result, to regulate the amount of atomic nitrog obtained. The mass of separated nitrogen obtained from the dissociation reaction determined by the relationship:

$$m_{N_2} = \frac{p}{1.6} \cdot s \cdot V_{in} \quad (5.14)$$

where p—pressure, s—volumetric proportion of ammonia dissociating in the inlet atmosphere. The mass of separated nitrogen obtained from the dissociation reacti in the function of time is determined by the relationship:

$$m_{N_2}(t) = \frac{p}{1.6} \cdot s \cdot F_{in}(t) \quad (5.15)$$

where F_{in} is the flow rate of the inlet atmosphere.

Controlling the nitriding process consists of adjusting the relationship between the flux of nitrogen to the steel surface and the flux of nitrogen from the surfa deep into the steel. This allows to control the growth kinetics of the nitrided lay The magnitude of the first flux is a function of nitrogen availability m_{N_2}, while the magnitude of the second flux depends on the content of alloying elements the nitrided steel. In the processes implemented in nitriding atmospheres obtaine from a single-component inlet atmosphere of ammonia (100% NH_3) and a two-component atmosphere with dissociated ammonia (NH_3/NH_{3diss}), there is a complete correspondence between the value of the potential and the availability of nitroge (an increase in the value of the potential is always accompanied by an increase the availability of nitrogen and vice versa). In nitriding atmospheres obtained from a two-component inlet atmosphere with nitrogen (NH_3/N_2) as the nitrogen content in the inlet atmosphere changes, the availability of nitrogen changes. However, does not need to be accompanied by a change in the value of the nitrogen pc nor a change in nitrogen potential does not have to be accompanied by a cha nitrogen availability [28, 68]. For a single-component inlet atmosphere of ammonia (NH_3), nitrogen availability is equal to the mass of nitrogen. For a two-compon inlet atmosphere of nitrogen diluted in the ratio of NH_3/N_2, the total mass of nitrogen is a higher value than the nitrogen availability, because the equatio nitrogen availability is based on the nitrogen from the dissociation of ammonia.

In conclusion, nitrogen availability ($m_{N_2}(t)$) is a parameter that relates the rate of ammonia dissociation to a technological parameter, i.e. the flow rate of the atmosphere, and contains information about the amount of nitrogen obtained du the nitriding process for a given rate of ammonia dissociation and a given flow of the inlet atmosphere [28].

By substituting the transformed expression (5.8) into the Eq. (5.14), the availability of nitrogen can be described as a function of the rate of dissociation of ammoni the flow rate of the inlet atmosphere:

$$m_{N_2}(t) = \frac{p}{1.6} \cdot \frac{}{2\ \check{S}} \cdot F_i(t) \tag{5.16}$$

The formula (5.14) exhibits that, at a constant-flow rate of inlet atmosphere ar pressure of nitriding atmosphere, nitrogen availability is a function of the rate ammonia dissociation and the pressure of the nitriding atmosphere.

To summarise the above considerations, nitrogen potential is an effective con parameter for the nitriding process only for nitriding atmospheres obtained fro single-component ammonia and those obtained from a two-component atmosph of ammonia and dissociated ammonia NH_3/NH_{3diss} in conditions of atmospheric pressure. For processes carried out in nitriding atmospheres obtained from a component atmosphere with nitrogen and implemented at a reduced pressure, is no direct relationship between the value of nitrogen potential and the compos of the inlet atmosphere. For such cases, the parameter describing the process rate of ammonia dissociation. As argued by J. Michalski and E. Wołowiec-Korec there are multiple compositions of atmosphere in a nitriding atmosphere obtair from a two-component inlet atmosphere with nitrogen, under conditions of varyi pressure, which are characterised by the same value of nitrogen potential and of ammonia dissociation for different pressures [27]. This characteristic can be used in the search for optimum values of pressures and compositions of inlet atmosph

Nitrogen potential as a function of rate of ammonia dissociation

It has been shown that the nitrogen potential can only be effectively used as a parameter for the nitriding process in nitriding atmospheres obtained from a singl component ammonia atmosphere or from a two-component mixture of ammonia dissociated ammonia at atmospheric pressure. It was also found that there is no relationship between the nitrogen potential and the composition of the input mixture in the nitriding process under low-pressure conditions in the atmosphere of the produced two-component mixture with nitrogen. In this situation, dependencies were established allowing to apply both the nitrogen potential and the ammonia dissociation rate for the selected ammonia pressure and mixture composition aNH_3/bN_2. This allows to analyse the equilibrium diagrams of Fe-N phases in the coordinate system such as *rate of ammonia dissociation—temperature* (ω,T) instead of the traditional *nitrogen potential—temperature* (N_P,T).

When considering the nitrogen potential of the nitriding atmosphere created by the incoming single-component ammonia, this can be determined as a function of the rate of dissociation of ammonia at atmospheric and reduced pressure from the following equations:

$$N_P(\omega) = \frac{1-\omega}{\left(\frac{3}{4}\cdot\omega\right)^{1.5}}, \qquad p_{NH_3} = 1013\,\text{hPa} \tag{5.17}$$

$$N_P(\omega,\, p_{NH_3}) = \frac{1-\omega}{\left(\frac{3}{4}\cdot\omega\right)^{1.5}\cdot\left(p_{NH_3}\right)^{0.5}}, \qquad p_{NH_3} < 1013\,\text{hPa} \tag{5.18}$$

When the inlet atmosphere is ammonia in the conditions of atmospheric pressure, the value of which does not change, the nitrogen potential is described by the relationship (Eq. 5.17) and only one value of the rate of dissociation corresponds to one value of potential. If a single-component atmosphere is used in a process in which the pressure in the furnace chamber is also a variable parameter, then the nitrogen potential is described by the equation (Eq. 5.18). In this case, the nitrogen potential is a function of two variables, and the graphical form of the function is the surface $N_P(\omega,\, p_{NH_3})$ in the coordinate system of the *rate of dissociation-pressure-nitrogen potential* (ω, p_{NH_3}, N_P) (Fig. 5.4). As seen from the figure, different pressures and rates of dissociation can correspond to the same value of nitrogen potential (line I) and different pressures and nitrogen potentials can correspond to the same rate of ammonia dissociation (line II).

The relationships (5.19) and (5.20) permit to calculate the nitrogen potential of the nitriding atmosphere obtained from a two-component inlet atmosphere with nitrogen at a proportion aNH_3/bN_2 as a function of rate of ammonia dissociation, respectively for atmospheric pressure and reduced pressure:

$$N_P(\omega, b) = \frac{1-b-\frac{1}{2}\cdot\omega\cdot(2-b)}{\left(\frac{3}{4}\cdot\omega\right)^{1.5}} \tag{5.19}$$

$$N_P(\omega, b,\, p_{aNH_3/bN_2}) = \frac{1-b-\frac{1}{2}\cdot\omega\cdot(2-b)}{\left(\frac{3}{4}\cdot\omega\right)^{1.5}\cdot\left(p_{aNH_3/bN_2}\right)^{0.5}} \tag{5.20}$$

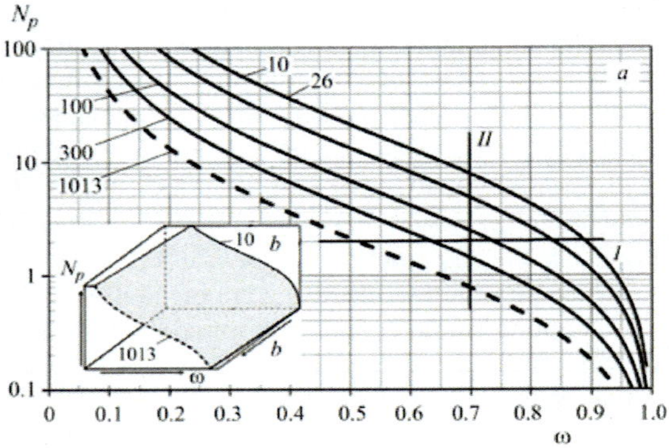

Fig. 5.4 Curves of nitrogen potential N_P as a function of rate of ammonia dissociation ω and ammonia pressure p_{NH_3}, forming the surface $N_P(\omega, p_{NH_3})$ described by Eq. (5.18): the numbers near the curves mark the level of pressure (hPa). Reprinted with permission from Ref. [27] Copyright (2019) Springer

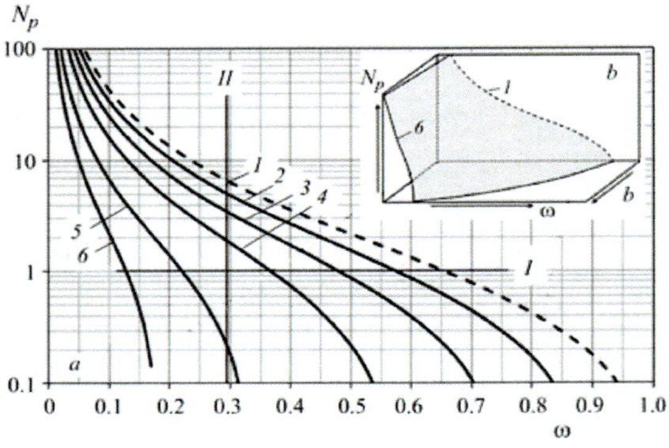

Fig. 5.5 Curves of nitrogen potential N_P as a function of the rate of dissociation of ammonia and the nitrogen content in the inlet atmosphere $a\mathrm{NH_3}/b\mathrm{N_2}$ forming the surface $N_P(\omega, b)$ described by Eq. (5.19): (1) b = 0, a = 1; (2) b = 0.2, a = 0.8; (3) b = 0.4, a = 0.6; (4) b = 0.6, a = 0.4; (5) b = 0.8, a = 0.2; (6) b = 0.9, a = 0.1. Reprinted with permission from Ref. [27] Copyright (2019) Springer

If the inlet atmosphere is a two-component atmosphere with nitrogen, then in the conditions of atmospheric pressure, the nitrogen potential is a function of the rate of dissociation and of nitrogen content in the inlet atmosphere (5.19). The graphical form of the function is the surface $N_P(\omega, b)$ in the coordinate system, the *rate of dissociation-nitrogen content in the inlet atmosphere-nitrogen potential* (ω,b,N_P). As can be seen from Fig. 5.5, in contrast to a single-component inlet atmosphere of ammonia, different contents of nitrogen in a two-component inlet atmosphere of ammonia-nitrogen can correspond to one value of the potential and to different values of the ammonia dissociation rate (line I) and one value of the ammonia dissociation rate can correspond to different values of the nitrogen potential (line II).

If a two-component atmosphere with nitrogen ($\mathrm{NH_3/N_2}$) is used in a process in which the pressure is also a variable parameter, then the nitrogen potential is described by Eq. (5.20). In this case, the nitrogen potential is a function of three variables, and the graphical form of the function is a solid limited by the surface $N_P(\omega, b, 1013\ \text{hPa})$ and $N_P(\omega, b, 26\ \text{hPa})$ and by the planes of the coordinate system (ω,b,N_P) (Fig. 5.6). There are multiple compositions of atmosphere $a\mathrm{NH_3}/b\mathrm{N_2}$ in a nitriding atmosphere

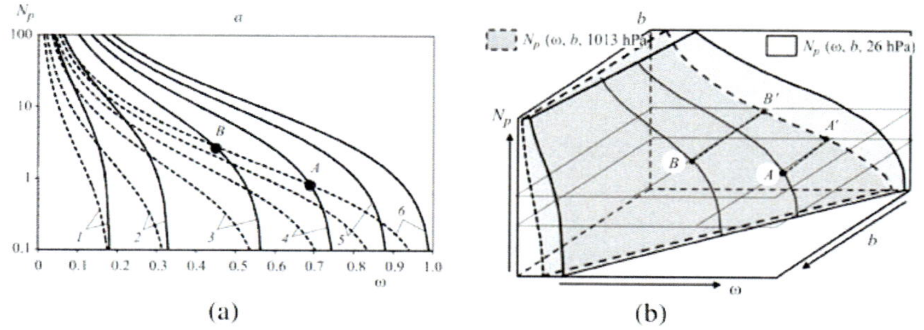

Fig. 5.6 Plots of the nitrogen potential as a function of the rate of dissociation of ammonia, nitrogen content and the pressure of nitriding atmosphere, making up the solid $N_P(\omega, b, p_{aNH_3/bN_2})$ described by Eq. (5.20), limited with the surfaces: $N_P(\omega,b$, 1013 hPa), $N_P(\omega, b$, 26 hPa) and planes of the coordinate system (ω, b, N_P). Reprinted with permission from Ref. [27] Copyright (2019) Springer

obtained from a two-component inlet atmosphere with nitrogen, under conditions of varying pressure, which are characterised by the same value of nitrogen potential and rate of ammonia dissociation for different pressure of the nitriding atmosphere. In Fig. 5.6b A-A', B-B' sections are marked, created from intersecting the surface $N_P(\omega,b$, 26 hPa) and the surface $N_P(\omega, b$, 1013 hPa) by a plane parallel to the plane $(\omega–b)$. Points A and B belong to the surface $N_P(\omega, b$, 26 hPa), while points 1' and 2' belong to the surface $N_P(\omega, b$, 26 hPa). Sections 1-1' and 2-2' lie inside the solid, and the nitrogen potentials lying on sections A-A' and B-B' satisfy the conditions $N_P = 0.8$, $\omega = 0.69$ and $N_P = 3.0$, $\omega = 0.44$, respectively.

Rate of dissociation of ammonia as a function of nitrogen potential

In the conditions of atmospheric pressure, the value of which is constant during the process, only one value of the nitrogen potential corresponds to one value of the rate of dissociation. The relationship describing the rate of dissociation of ammonia as a function of nitrogen potential is the solution of Eq. (5.17) relative to N_P:

$$\omega(N_P) \approx 0.970256 \cdot (N_P + 0.98788404)^{-0.59265012}, \qquad N_P > 0 \tag{5.21}$$

By substituting in place of the nitrogen potential N_P its limit values $N_P^{\alpha/\gamma'}$ [atm.$^{-0.5}$] and $N_P^{\gamma'/\varepsilon}$ [atm.$^{-0.5}$] corresponding to the thermodynamic stability of the nitride phases (dependencies (5.22) and (5.23)), the functions describing the temperature-dependent boundary rates of ammonia dissociation $\omega^{\alpha/\gamma'}$ and $\omega^{\gamma'/\varepsilon}$ are obtained, which correspond to limit values of nitrogen potentials $N_P^{\alpha/\gamma'}(T)$ and $N_P^{\gamma'/\varepsilon}(T)$ of E. Lehrer's system (dependencies (5.24) and (5.25)).

$$N_P^{\alpha/\gamma'} = 318.2766 \cdot \exp\left(\frac{4556}{T} - 12.88\right) \tag{5.22}$$

$$N_P^{\gamma'/\varepsilon} = 318.2766 \cdot \exp\left(\left(\frac{60536}{T} - 56.85\right)^{0.5} - 9.63\right) \tag{5.23}$$

$$\omega^{\alpha/\gamma'}(T) = 0.70256 \cdot \left(N_P^{\alpha/\gamma'}(T) + 0.98799404\right)^{-0.59265012} \tag{5.24}$$

$$\omega^{\gamma'/\varepsilon}(T) = 0.970256 \cdot \left(N_P^{\gamma'/\varepsilon}(T) + 0.98799404\right)^{-0.59265012} \tag{5.25}$$

Graphical representation of the functions (5.24) and (5.25) is shown in Fig. (5.7b). It is the system of T–ω, which is a transformation of E. Lehrer system of T–N_P. The boundary rates of ammonia dissociation $\omega^{\alpha/\gamma'}(T)$, $\omega^{\gamma'/\varepsilon}(T)$ describe monovalent functions, similar to the function describing the nitrogen potential for a single-component inlet atmosphere of ammonia. It is seen from the graph that with the growing temperature, the phase stability area ε is rising, while the phase stability area α changes. The phase stability area γ' changes slightly.

If the nitriding process is carried out in a nitriding atmosphere obtained from a single-component inlet atmosphere of ammonia under conditions of varying pressure, the value of the nitrogen potential is a function of the rate of dissociation and pressure (Eq. 5.18). A solution of this equation relative to the rate of dissociation is a function of two variables, i.e. nitrogen potential and pressure. The coefficients u, v, w of Eq. (5.26) are a function of pressure as shown in Table 5.1. Thus:

$$\omega\left(N_P, p_{NH_3}\right) \approx u\left(p_{NH_3}\right) \cdot \left[N_P - v\left(p_{NH_3}\right)\right]^{w(p_{NH_3})}, \qquad \omega > 0 \tag{5.26}$$

A graphical form of the function is the surface $\omega(N_P, p_{NH_3})$ in the coordinate system *nitrogen potential-pressure-rate of dissociation* (N_P,p_{NH_3},ω) (Fig. 5.8). In contrast to a single-component inlet atmosphere of ammonia at atmospheric pressure, constant values of nitrogen potential correspond to different values of dissociation rates (line I), while constant values of ammonia dissociation rates correspond to different values of nitrogen potentials (line II).

The functions describing the boundary rates of ammonia dissociation $\omega^{\alpha/\gamma'}$ (Eq. 5.27) and $\omega^{\gamma'/\varepsilon}$ (Eq. 5.28) depending on temperature, which correspond to boundary nitrogen potentials $N_P^{\alpha/\gamma'}(T)$ and $N_P^{\gamma'/\varepsilon}(T)$ of E. Lehrer's system, are obtained by substituting the values of the boundary potentials as a function of temperature $N_P^{\alpha/\gamma'}(T)$ and $N_P^{\gamma'/\varepsilon}(T)$ in Eq. (5.26).

$$\omega^{\alpha/\gamma'}\left(T, p_{NH_3}\right) \approx u\left(p_{NH_3}\right) \cdot \left[N_P^{\alpha/\gamma'}(T) - v\left(p_{NH_3}\right)\right]^{w(p_{NH_3})} \tag{5.27}$$

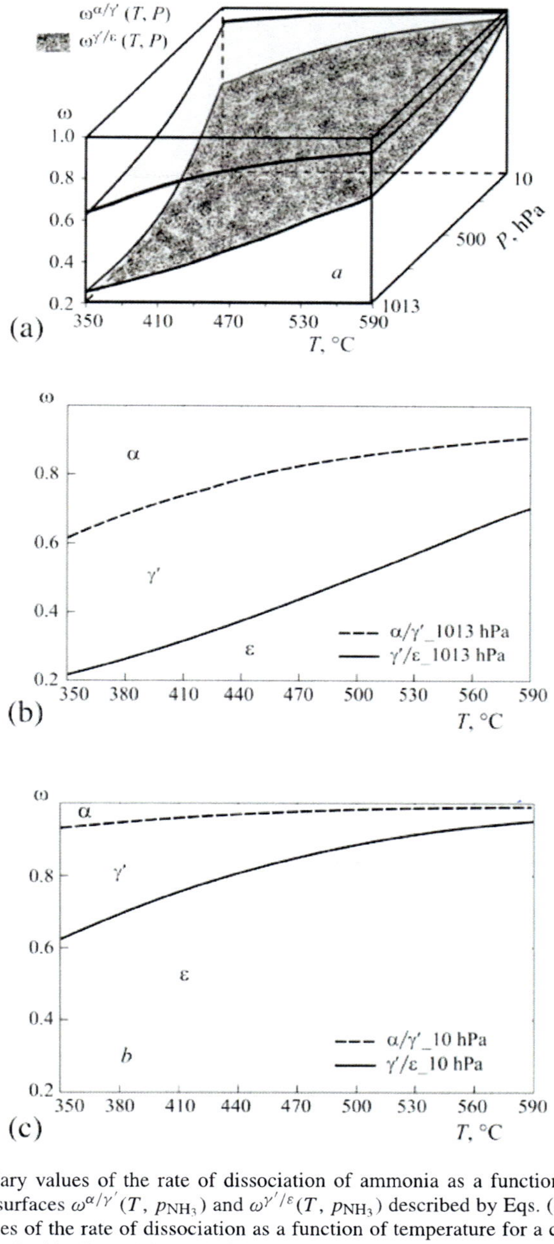

Fig. 5.7 Boundary values of the rate of dissociation of ammonia as a function of temperature and pressure: **a** surfaces $\omega^{\alpha/\gamma'}(T, p_{NH_3})$ and $\omega^{\gamma'/\varepsilon}(T, p_{NH_3})$ described by Eqs. (5.27) and (5.28), **b** boundary values of the rate of dissociation as a function of temperature for a constant pressure of 1013 hPa described by Eqs. (5.24) and (5.25), **c** boundary values of the rate of dissociation as a function of temperature for a constant pressure of 10 hPa. Reprinted with permission from Ref. [27] Copyright (2019) Springer

Table 5.1 Coefficients of Eq. (5.26). Reprinted with permission from Ref. [27] Copyright (2019) Springer

p_{NH_3} [hPa]	u	v	w
10	3.87851	–9.94485	–0.59265
26	2.922182	–6.16777	–0.59265
50	2.407363	–4.44746	–0.59265
100	1.960372	–3.14483	–0.59265
200	1.596376	–2.22373	–0.59265
300	1.415647	–1.81567	–0.59265
400	1.299967	–1.57241	–0.59265
500	1.216789	–1.40641	–0.59265
600	1.152794	–1.28387	–0.59265
700	1.101321	–1.18863	–0.59265
800	1.058594	–1.11186	–0.59265
900	1.022284	–1.04827	–0.59265
1013	0.986988	–0.98793	–0.59265

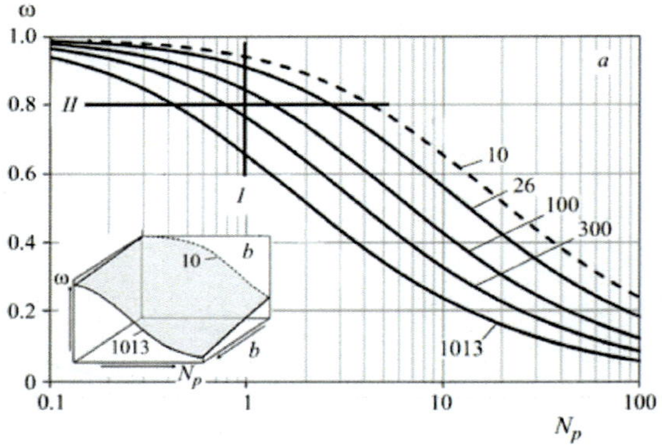

Fig. 5.8 Plots of the rate of dissociation of ammonia $\omega(N_P,p_{NH_3})$ as functions of the nitrogen potential N_P and the pressure of ammonia p_{NH_3} (the numbers near the curves mark the level of pressure in hPa) **a**, which form the surface $\omega(N_P,p_{NH_3})$ described by relation (5.26) **b**. Reprinted with permission from Ref. [27] Copyright (2019) Springer

$$\omega^{\gamma'/\varepsilon}\left(T, p_{NH_3}\right) \approx u\left(p_{NH_3}\right) \cdot \left[N_P^{\gamma'/\varepsilon}(T) - v\left(p_{NH_3}\right)\right]^{w(p_{NH_3})} \qquad (5.28)$$

The graphical forms of the function (5.27) and (5.28) are the surfaces of the boundary rates of ammonia dissociation $\omega^{\alpha/\gamma'}$ and $\omega^{\gamma'/\varepsilon}$ in the coordinate system: *temperature-pressure-rate of dissociation* of nitriding atmosphere (T,p_{NH_3},ω) (Fig. 5.7a). Figure 5.7c shows the values of the boundary rates of dissociation $\omega^{\alpha/\gamma'}$ and $\omega^{\gamma'/\varepsilon}$ for the pressure of 10 hPa. It is seen when comparing Fig. 5.7b, c that along with the declining pressure, the stability area of the phase α and phase γ' is lowering and the stability area of the phase ε is rising.

Solving Eq. (5.19) against ω creates a dependence of the nitrogen potential from the rate of dissociation of ammonia and from the nitrogen content in the inlet atmosphere. The general form of this function is the same as for the single-component

Table 5.2 Values of the coefficients of Eq. (5.29) for selected compositions of inlet atmosphere aNH_3/bN_2. Reprinted with permission from Ref. [27] Copyright (2019) Springer

a	b	u	v	w	N_P^{max}	N_P^{min}
0.9	0.1	0.921263	–0.96320	–0.59244	1371	0.1
0.8	0.2	0.853418	–0.94287	–0.59260	1218	0.1
0.7	0.3	0.780883	–0.92255	–0.59210	1065	0.1
0.6	0.4	0.706516	–0.91338	–0.59279	911	0.1
0.5	0.5	0.625405	–0.90650	–0.59244	758	0.1
0.4	0.6	0.537712	–0.91022	–0.59173	605	0.1
0.3	0.7	0.442678	–0.93994	–0.59162	452	0.1
0.2	0.8	0.335903	–1.02210	–0.59181	299	0.1
0.1	0.9	0.207555	–1.26086	–0.59058	145	0.1

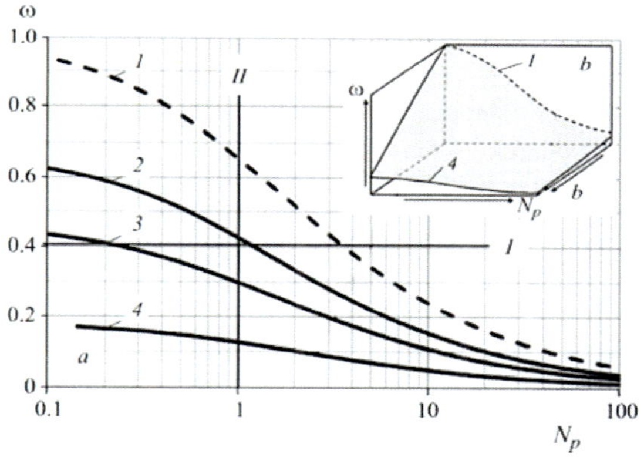

Fig. 5.9 Plots of the rate of dissociation of ammonia as functions of the nitrogen potential and the nitrogen content in the inlet atmosphere aNH_3/bN_2, forming the surface $\omega(N_P,b)$ determined by Eq. (5.29): (1) b = 0, a = 1; (2) b = 0.5, a = 0.5; (3) b = 0.7, a = 0.3; (4) b = 0.9, a = 0.1. Reprinted with permission from Ref. [27] Copyright (2019) Springer

inlet atmosphere (5.26) except that the coefficients u, v, w of Eq. (5.29) are a function of nitrogen content (Table 5.2):

$$\omega(N_P, b) \approx u(b)[N_P - v(b)]^{w(b)} \tag{5.29}$$

where $u(b)$, $v(b)$, $w(b)$—coefficients of an approximation function with the form $y = u(x - v)^w$, b—nitrogen content in a two-component inlet atmosphere.

The graphical form of the function described with Eq. (5.29) is the surface $\omega(N_P,b)$ in the coordinate system *nitrogen potential-nitrogen content in the inlet atmosphere-dissociation rate* (N_P,b,ω) (Fig. 5.9). In contrast to a single-component inlet atmosphere of ammonia at atmospheric pressure, for different nitrogen contents in the inlet atmosphere, different values of nitrogen potential and nitrogen content in the inlet atmosphere can correspond to the same value of the rate of dissociation. In addition, different values for the rates of dissociation of ammonia and of nitrogen content in the inlet atmosphere may correspond to the same value of nitrogen potential.

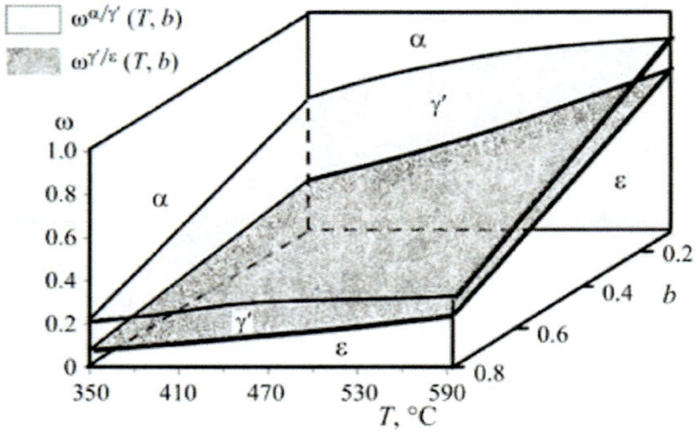

Fig. 5.10 Limit degrees of dissociation of ammonia as functions of temperature and the nitrogen content in the input atmosphere $a\mathrm{NH_3}/b\mathrm{N_2}$, forming the surfaces: $\omega^{\alpha/\gamma'}(T,b)$ and $\omega^{\gamma'/\varepsilon}(T,b)$, determined by Eqs. (5.30) and (5.31). Reprinted with permission from Ref. [27] Copyright (2019) Springer

By substituting to Eq. (5.29) the relationships (5.22) and (5.23), the functions for calculating the boundary values of the rates of ammonia dissociation $\omega^{\alpha/\gamma'}$ and $\omega^{\gamma'/\varepsilon}$ are obtained. These functions depend on the temperature for specific compositions of a two-component inlet atmosphere with nitrogen (Fig. 5.10), which correspond to the boundary values of nitrogen potentials $N_P^{\alpha/\gamma'}(T)$ and $N_P^{\gamma'/\varepsilon}(T)$ of the E. Lehrer system (Eqs. 5.30 and 5.31). As the content of nitrogen in the inlet atmosphere is rising, the stability area of the phase ε and the phase γ' decreases, while the stability area of the α phase increases.

$$\omega^{\alpha/\gamma'}(T,b) \approx u(b)\cdot\left[N_P^{\alpha/\gamma'}(T) - v(b)\right]^{w(b)} \tag{5.30}$$

$$\omega^{\gamma'/\varepsilon}(T,b) \approx u(b)\cdot\left[N_P^{\gamma'/\varepsilon}(T) - v(b)\right]^{w(b)} \tag{5.31}$$

In the case of a nitriding atmosphere obtained from a two-component inlet atmosphere, taking into account the pressure changes, the potential values are described by Eq. (5.30). By solving it, a function of three variables is obtained, i.e. nitrogen potential, nitrogen content in the inlet atmosphere and pressure:

$$\omega(N_P, b, p_{a\mathrm{NH_3}/b\mathrm{N_2}}) \approx u(b, p_{a\mathrm{NH_3}/b\mathrm{N_2}})\cdot\left[N_P - v(b, p_{a\mathrm{NH_3}/b\mathrm{N_2}})\right]^{w(b, p_{a\mathrm{NH_3}/b\mathrm{N_2}})} \tag{5.32}$$

where u(b,$p_{a\mathrm{NH_3}/b\mathrm{N_2}}$), v(b,$p_{a\mathrm{NH_3}/b\mathrm{N_2}}$), w(b, $p_{a\mathrm{NH_3}/b\mathrm{N_2}}$)—coefficients of an approximation function $y = u\cdot(x-v)^w$, b—nitrogen content in a two-component inlet atmosphere.

There are multiple compositions of atmosphere $a\mathrm{NH_3}/b\mathrm{N_2}$ in a nitriding atmosphere, obtained from a two-component inlet atmosphere with nitrogen, under conditions of varying pressure, which are characterised by the same value of the rate of ammonia dissociation and nitrogen potential for different pressures of the atmosphere. This characteristic is useful in the search for equivalent atmospheres in terms of the value of the rate of dissociation of ammonia in the aspect of nitrogen availability of the nitriding atmosphere for an assumed value of nitrogen potential, particularly in nitrogen-diluted atmospheres.

By substituting to Eq. (5.32) the values of the boundary potentials as a function of temperature $N_P^{\alpha/\gamma'}(T)$ and $N_P^{\gamma'/\varepsilon}(T)$, the functions are obtained describing the boundary rates of ammonia dissociation $\omega^{\alpha/\gamma'}$ and $\omega^{\gamma'/\varepsilon}$ depending on temperature, nitrogen content in the inlet atmosphere and the pressure of nitriding atmosphere ($p_{a\mathrm{NH_3}/b\mathrm{N_2}}$):

$$\omega^{\alpha/\gamma'}(N_P^{\alpha/\gamma'}, b, p_{a\mathrm{NH_3}/b\mathrm{N_2}}) \approx u(b, p_{a\mathrm{NH_3}/b\mathrm{N_2}}) \cdot \cdot\left[N_P^{\alpha/\gamma'}(T) - v(b, p_{a\mathrm{NH_3}/b\mathrm{N_2}})\right]^{w(b, p_{a\mathrm{NH_3}/b\mathrm{N_2}})} \tag{5.33}$$

$$\omega^{\gamma'/\varepsilon}(N_P^{\gamma'/\varepsilon}, b, p_{a\mathrm{NH_3}/b\mathrm{N_2}}) \approx u(b, p_{a\mathrm{NH_3}/b\mathrm{N_2}}) \cdot \cdot\left[N_P^{\gamma'/\varepsilon}(T) - v(b, p_{a\mathrm{NH_3}/b\mathrm{N_2}})\right]^{w(b, p_{a\mathrm{NH_3}/b\mathrm{N_2}})} \tag{5.34}$$

The functions (5.33) and (5.34) correspond to the boundary values of the nitrogen potentials $N_P^{\alpha/\gamma'}(T)$ and $N_P^{\gamma'/\varepsilon}(T)$ of the E. Lehrer system. The graphical form of the function (5.33) and (5.34) are solids limited with the surfaces $\omega^{\alpha/\gamma'}(T,b,p_{a\mathrm{NH_3}/b\mathrm{N_2}})$ (Fig. 5.11a) and surfaces $\omega^{\gamma'/\varepsilon}(T,b,p_{a\mathrm{NH_3}/b\mathrm{N_2}})$ (Fig. 5.11b) (pressures of 1013 hPa and 26 hPa are assumed for illustrations), which correspond to the boundary nitrogen potentials $N_P^{\alpha/\gamma'}(T)$ and $N_P^{\gamma'/\varepsilon}(T)$ of the E. Lehrer system. As can be seen from Fig. 5.11, the boundary values of the dissociation rates change to a greater extent as a result of change in the composition of the inlet atmosphere than as a result of change in pressure.

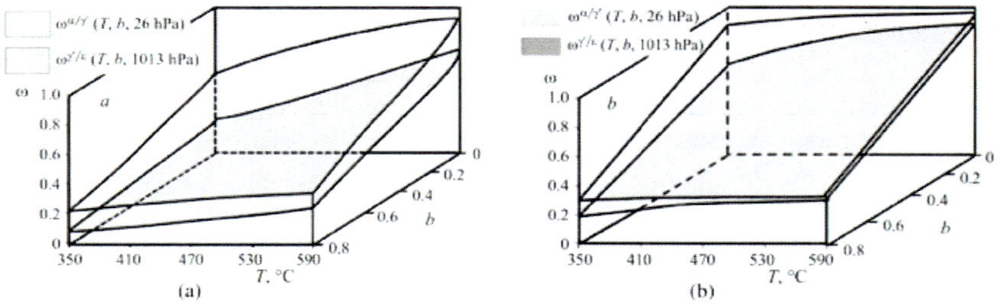

Fig. 5.11 Plots of the degree of dissociation of ammonia as functions of temperature, nitrogen content and pressure of nitriding atmosphere, forming a solid described by the function $\omega^{\alpha/\gamma'}(T, b, p_{a\mathrm{NH_3}/b\mathrm{N_2}})$ (Eq. 5.33) **a**, and $\omega^{\gamma'/\varepsilon}(T, b, p_{a\mathrm{NH_3}/b\mathrm{N_2}})$ Eq. (5.34) **b** limited with the surfaces of the boundary degrees of dissociation for the pressure of 26 and 1013 hPa. Reprinted with permission from Ref. [27] Copyright (2019) Springer

It should be concluded when summarising the impact of the nitriding atmosphere on the character of the nitrided layer that for a single-component inlet nitriding atmosphere (100% NH_3), as the temperature grows in the structure of the nitrided layer, the stability area of the ε phase grows, while the stability area of the α phase lowers. The phase stability area γ' changes slightly. Along with the decreasing pressure, the stability area of the phase α and phase γ' decreases and the stability area of the phase ε increases. For two-component inlet nitriding atmospheres, as the degree of dilution of the inlet atmosphere with nitrogen increases, irrespective of the pressure, the stability area of the α phase increases, while the stability areas of the γ' phase and the ε phase are reduced. For both, atmospheric pressure and reduced pressure, by changing the nitrogen content in the inlet atmosphere, the rate of ammonia dissociation can be altered over a wide range of values. As the degree of dilution of the inlet atmosphere with nitrogen rises, irrespective of the pressure, the stability area of the α phase rises, while the stability areas of the γ' phase and the ε phase are reduced.

In the case of a nitriding atmosphere obtained from a single-component inlet atmosphere of ammonia in conditions of atmospheric pressure (1013 hPa), the value of the nitrogen potential and the rate of ammonia dissociation equally define the nitrogen flux. This is due to the fact that one value of nitrogen potential corresponds to one value of the rate of ammonia dissociation, as indicated by Eqs. (5.17) and (5.21). A change in the value of the nitrogen potential/of the rate of ammonia dissociation can be achieved by varying the ammonia flow rate. A change in the value of the nitrogen potential/of the rate of ammonia dissociation at a constant-flow rate of ammonia can be achieved by varying the pressure of the nitriding atmosphere, as results from Eqs. (5.18), (5.25) and Figs. (5.5) and (5.8). This shows that in a nitriding atmosphere produced from a single-component inlet atmosphere, pressure is an effective parameter enabling to alter the rate of ammonia dissociation and, consequently, nitrogen availability. In a nitriding atmosphere obtained from a two-component inlet atmosphere with nitrogen, the rate of ammonia dissociation, at a constant-flow rate of the inlet atmosphere, can be changed by changing the composition of the inlet atmosphere or by changing the pressure. Changing the composition of the inlet atmosphere changes the rate of dissociation to a greater extent than changing the pressure (Fig. 5.11).

The common parameter characterising the processes described above is the rate of dissociation of ammonia, while the nitrogen potential is only applicable to some of them.

Control of nitrogen flux in nitriding process

Efforts should be made in the nitriding process to match the nitrogen flux from the nitriding atmosphere to the mass of nitrogen used to constitute the nitrided layer. The nitrogen mass of the nitrided layer (m_{NL}) is the sum of the mass of nitrogen in the layer of iron nitrides ($m_{\gamma'+\varepsilon}$) and the mass of nitrogen in the solution layer (m_α):

$$m_{NL} = m_{\gamma'+\varepsilon} + m_\alpha \tag{5.35}$$

The kinetics of the growth of the nitrided layer, and hence of its mass increase, follows a parabolic law for both the near-surface layer of iron nitrides and the solution layer

$$m_{NL} = (k_{\gamma'+\varepsilon} + k_{\alpha}) \cdot \sqrt{t} \tag{5.36}$$

where $k_{\gamma'+\varepsilon}$—a parabolic constant of growth of a layer of iron nitrides, k_{α}—a parabolic constant of growth of a solution layer, t—duration. The absorption capacity of the nitrided steel as a function of time is determined by the derivative of the function (Eq. 5.36) with respect to time:

$$\frac{\partial(m_{NL})}{\partial t} = \frac{k_{\gamma'+\varepsilon} + k_{\alpha}}{2 \cdot \sqrt{t}} \tag{5.37}$$

To obtain a nitrided layer of m_{NL}, the nitrogen availability must meet the following conditions:

$$\beta_N \cdot m_{N_2} \geq \frac{\partial(m_{NL})}{\partial t} \tag{5.38}$$

where β_N—nitrogen utilisation coefficient, m_{N_2}—nitrogen availability in the nitriding atmosphere.

The relationship (5.39) resulting from Eq. (5.38) can be used to design gas nitriding processes in the conditions of atmospheric pressure as well as at reduced pressure:

$$\beta_N \cdot \frac{p_{NH_3}}{1.6} \cdot \frac{\omega}{2-\omega} \cdot F_{in} \geq \frac{k_{\gamma'+\varepsilon} + k_{\alpha}}{2 \cdot \sqrt{t}} \tag{5.39}$$

where F_{in}—flow rate of the inlet atmosphere.

Depending on the process variant resulting from the type of inlet atmosphere used and the available control parameters, the left-hand side of Eq. (5.39) will be modified. The parameter permitting to change in the rate of ammonia dissociation is pressure in the case of a single-component atmosphere for a constant-flow rate of the inlet atmosphere. After substituting the relationship (5.26) to Eq. (5.39) in place of ω, the inequality (5.39) takes the form:

$$\beta_N \cdot \frac{p_{NH_3}}{1.6} \cdot \frac{u(p_{NH_3}) \cdot [N_P - v(p_{NH_3})]^{w(p_{NH_3})}}{2 - u(p_{NH_3}) \cdot [N_P - v(p_{NH_3})]^{w(p_{NH_3})}} \cdot F_{in} \geq \frac{k_{\gamma'+\varepsilon} + k_{\alpha}}{2 \cdot \sqrt{t}} \tag{5.40}$$

In the case of a nitriding atmosphere obtained from a two-component inlet atmosphere with nitrogen, both the pressure and the nitrogen content in the inlet atmosphere enable to change in the rate of ammonia dissociation, but most vacuum treatment equipment operates at a constant vacuum. In this case, the inequality (5.39), after substituting Eq. (5.32), takes the form

$$\beta_N \cdot \frac{p_{a\mathrm{NH_3}/b\mathrm{N_2}}}{1.6} \cdot \frac{u(b,\, p_{a\mathrm{NH_3}/b\mathrm{N_2}}) \cdot [N_P - v(b,\, p_{a\mathrm{NH_3}/b\mathrm{N_2}})]^{w(b,\, p_{a\mathrm{NH_3}/b\mathrm{N_2}})}}{2 - u(b,\, p_{a\mathrm{NH_3}/b\mathrm{N_2}}) \cdot [N_P - v(b,\, p_{a\mathrm{NH_3}/b\mathrm{N_2}})]^{w(b,\, p_{a\mathrm{NH_3}/b\mathrm{N_2}})}} \cdot F_{\mathrm{in}} \geq \frac{k_{\gamma'+\varepsilon} + k_\alpha}{2 \cdot \sqrt{t}}, \quad p_{a\mathrm{NH_3}/b\mathrm{N_2}} = \mathrm{const.} \tag{5.41}$$

The proposed method of controlling and regulating the nitrogen flux to the nitrided steel surface is an alternative to the solutions currently employed in nitriding processes, particularly low-pressure processes.

5.7 Further Directions of Development

Two types of transformations can take place in metals or their alloys subjected to the activity of high pressure: a shift in phase equilibrium temperatures even in the condensed (liquid or solid) state or allotropy (polymorphism). For low-pressure nitriding, the direction of action of the reduced pressure is reversed, which can manifest itself as losses in the nitrided layer when it is annealed under the conditions of technical vacuum [102]. A question remains open what is the form of the limit pressure function of the nitriding atmosphere $p(T)$, at which the nitrogen flux across the gas-surface boundary is 0? What is the limit pressure of the active nitriding atmosphere $p(T)$ that ensures the development of a solution layer only? It is seen by analysing the literature that the effect of pressure on the durability of the nitrided layer has not been well defined. It seems that these questions can be answered by way of numerical recalculations. Also, the development of unifying relationships for the description of gas nitriding described in this chapter, extended to other nitriding mixtures, including three-component mixtures, would contribute to the further development of knowledge on nitriding.

The state of understanding of the mechanisms determining the rate of nucleation and growth of the nitride zone during gas nitriding and carbonitriding has been expanded in recent years, but these works should continue until all gaps are clarified. M. Sommers, in his paper [103], identifies the following challenges in understanding further the development of the nitride and solution zone in terms of thermodynamics, diffusion kinetics and kinetics of surface reactions and enabling its mathematical modelling: (1) a systematic study of the thermodynamics of the Fe-N-C diagram and a description of this phase diagram, (2) a systematic study of the porosity in Fe-N(-C) system phases, including both homogeneous phases of homogeneous composition and materials with a complex layer, (3) determination of the reaction rates of the relevant surface reactions, (4) determination of N and C diffusion coefficients in single-phase samples in order to obtain a set of diffusion coefficients independent of the applied layer growth model, (5) systematic investigation of the role of residual stresses that arise in the nitrided layer (compound layer), including their influence on thermodynamics and kinetics, as well as the development of a model for their mathematical prediction. These issues are mostly still awaiting clarification.

References

1. G. Totten, Heat treating in 2020: What are the most critical issues and-What will the future look like? Heat Treat. Metals **31**(1), 1–3 (2004)
2. J. Michalski, J. Tacikowski, I. Sułkowski, P. Wach, S. Kopeć, Adjustable gas nitriding for use to highly loaded aerospace parts. Inżynieria Materiałowa **5**, 113–226 (2002). (in Polish)
3. D. Pye, *Practical Nitriding and Ferritic Nitrocarburizing* (ASM International, Materials Park, 2003)
4. J. Michalski, J. Tacikowski, P. Wach, J. Iwanow, M. Błasiak, M. Korwin, Anticorrosive nitriding of machine parts. Inżynieria Powierzchni **3**, 12–17 (2006). (in Polish)
5. R. Simons, Why nitriding steel is growing in popularity. Ind. Heat. **3** (2008)
6. M. Somers, E. Mittemeijer, Layer-growth kinetics on gaseous nitriding of pure iron: Evaluation of diffusion coefficients for nitrogen in iron nitrides. Metall. Mater. Trans. A **26**, 57–74 (1995)
7. N. Anichkina, V. Bogolyubov, V. Boiko, V. Denisov, I. Dukarevich, Comparison of methods of gas, ionic, and vacuum nitriding. Metal Sci. Heat Treat. **31**(3), 170–174 (1989)
8. M. Yang, R. Sisson, Alloy effects on the gas nitriding process. J. Mater. Eng. Perform. **23**(12), 4181–4186 (2014)
9. L. Barrallier, Classical nitriding of heat treatable steel, in *Thermochemical Surface Engineering of Steels* (Elsevier, 2015), pp. 393–412
10. M. Betiuk, J. Michalski, S. Kowalski, A. Nakonieczny, *Method of Glow Nitriding of Metal Objects*. Patent PL 217543 (Institute of Precision Mechanics, Poland, 2014)
11. K. Cho, K. Song, S.H. Oh, Y.-K. Lee, W. Lee, Enhanced surface hardening of AISI D2 steel by atomic attrition during ion nitriding. Surface Coat. Technol. **251**, 115–121 (2014)
12. D. Manova, D. Hirsch, J. Gerlach, S. Mändl, H. Neumann, B. Rauschenbach, In situ investigation of phase formation during low-energy ion nitriding of Ni80Cr20 alloy. Surf. Coat. Technol. **259**, 434–441 (2014)
13. I. Rosales, H. Martinez, R. Guardian, Mechanical performance of thermally post-treated ion-nitrided steels. Appl. Surf. Sci. **371**, 576–582 (2016)
14. D. Hoche, J. Kaspar, P. Schaaf, Laser nitriding and carburization of materials, in *Laser Surface Engineering*, eds. by J. Lawrence, C. Dowding, D. Waugh, J. Griffiths (Elsevier, 2015), pp. 33–58
15. Y. Lakhtin, Y. Kogan, S. Soshkin, Nitriding of steels in vacuum. Metal Sci. Heat Treat. **22**(9), 635–638 (1980)
16. S. Soshkin, Y. Lakhtin, Y. Kogan, Structure of the diffusion layer with vacuum nitriding. Metal Sci. Heat Treat. **26**(7), 521–523 (1984)
17. D. Jordan, H. Antes, V. Osterman, T. Jones, Low-torr range vacuum nitriding of 4140 steel. Heat Treat. Progr. **3–4**, 33–38 (2008)
18. D. Jordan, Vacuum gas nitriding furnace produces precision nitrided parts. Heat Treat. Progr. **9**, 45–48 (2009)
19. D. Jordan, T. Jones, H. Antes, No hydrogen embrittlement with low-pressure gas carburizing. Heat Treat. Progr. **3–4**, 29–31 (2009)
20. D. Jordan, *Controlling Compound (white) Layer Formation During Vacuum Gas Nitriding, Technical Report* (Solar Atmospheres Inc., Souderton, 2010)
21. P. Kula, R. Pietrasik, E. Wołowiec, B. Januszewicz, A. Rzepkowski, Low-pressure nitriding according to the FineLPN technology in multi-purpose vacuum furnaces. Adv. Mater. Res. **586**, 230–234 (2012)

22. M. Korecki, P. Kula, E. Wołowiec, M. Bazel, M. Sut, Low-pressure carburizing and nitriding of fuel injection nozzles. Heat Process. **3**, 59–62 (2014)
23. P. Kula, R. Pietrasik, E. Wołowiec, *Method of Nitriding Tools Made of Iron Alloys*. Patent PL 219125 (Lodz University of Technology, Poland, 2014)
24. Z. Zhou, M. Dai, Z. Shen, J. Hu, Effect of D.C. Electric field on salt bath nitriding for 35 steel and kinetics analysis. J. Alloys Comp. **623**, 261–265 (2015)
25. M. Perez, F. Belzunce, A comparative study of salt-bath nitrocarburizing and gas nitriding followed by post-oxidation used as surface treatments of H13 hot forging dies. Surf. Coat. Technol. **305**, 146–157 (2016)
26. J. Michalski, E. Wołowiec-Korecka, A study of the parameters of nitriding processes. Part 1. Metal Sci. Heat Treat. **61**(3–4), 183–190 (2019)
27. J. Michalski, E. Wołowiec-Korecka, A study of the parameters of nitriding processes. Part 2. Metal Sci. Heat Treat. **61**(5–6), 351–359 (2019)
28. J. Michalski, Using nitrogen availability as a nitriding process parameter. Ind. Heat. **10**, 63–68 (2012)
29. N. Kardonina, A. Yurovskikh, A. Kolpakov, Transformations in the Fe-N system. Metal Sci. Heat Treat. **52**(9–10), 457–467 (2010)
30. E. Wołowiec-Korecka, J. Michalski, B. Kucharska, Kinetic aspects of low-pressure nitriding process. Vacuum **155**, 292–299 (2018)
31. W. Arabczyk, J. Zamlynny, Study of the ammonia decomposition over iron catalysts. Catal. Lett. **60**(3), 167–171 (1999)
32. R. Pełka, *Studies on the Kinetics of the Process of Obtaining Nanocrystalline Iron Compounds FexO, FexN*. PhD thesis (West Pomeranian University of Technology in Szczecin, Szczecin, 2009). (in Polish)
33. E. Lehrer, über das Eisen-Wasserstoff-Amoniak-Gleichgewicht. Zeitschrift für Elektrochemie **36**(6), 383–392 (1930). (in German)
34. J. Kunze, *Nitrogen and Carbon in Iron and Steels-Thermodynamics* (Akademie Verlag, Berlin, 1990)
35. B. Kooi, M. Somers, E. Mittemeijer, An evaluation of the FeN phase diagram considering long range order of N atoms, γ'-Fe4N(1–x) and ε-Fe2N(1-z). Metall. Mater. Trans. A **27**, 1064–1071 (1996)
36. J. Jennings, *Catalytic Ammonia Synthesis Fundamentals and Practice* (Plenum Press, New York, 1991)
37. K. Aika, L. Christiansen, I. Dybkjaer, J. Hansen, P. Nielsen, A. Nielsen, P. Stolze, K. Tamaru, *Ammonia Catalysis and Manufacture* (Springer Verlag, Berlin/Heidelberg, 1995)
38. K. Jack, The occurrence and the crystal structure of α-iron nitride; A new type of interstitial alloy formed during the tempering of nitrogen-martensite. Proceed. R. Soc. Lond. **208**, 216–224 (1951)
39. K. Jack, Iron-nitrogen system: The crystal structures of ε-phase iron nitrides. Acta Crystall. **5**, 404–411 (1952)
40. L. Małdziński, *Thermodynamic, Kinetic and Technological Aspects of Producing Nitrided Layers on Iron and Steel in Processes of Gas Nitriding* (Poznan University of Technology, Poznan, 2002). (in Polish)
41. L. Małdziński, J. Tacikowski, Concept of an economical and ecological process of gas nitriding of steel. HTM Hartereitechnische Mitteilungen **61**(6), 296–302 (2006)
42. A. Yurovskikh, N. Kardonina, A. Kolpakov, Phase transformations in nitrided iron powders. Metal Sci. Heat Treat. **57**(7–8), 507–514 (2015)
43. I. Ozdemir, F. Akar, Local dynamics of chemical kinetics at different phases of nitriding process. J. Mater. Eng. Perform. **24**(8), 2984–2989 (2015)
44. S. Malinov, A. Böttger, E. Mittemeijer, M. Pekelharing, M. Somers, Phase transformations and phase equilibria in the Fe-N system at temperatures below 573K. Metall. Mater. Trans. A **32**(1), 59–73 (2001)
45. T. Liapina, A. Leineweber, E. Mittemeijer, Nitrogen redistribution in ε/γ'-iron nitride compound layers upon annealing. Scripta Materialia **48**(12), 1643–1648 (2003)

46. T. Liapina, *Phase Transformations in Interstitial Fe-N Alloys*. PhD thesis (Universitat Stuttgart, Stuttgart, 2005)
47. T. Liapina, A. Leineweber, E. Mittemeijer, Phase transformations in ε-/γ'-iron nitride compound layers in the temperature range of 613K–693K. Defect Diffusion Forum **237–240**, 1147–1152 (2005)
48. T. Liapina, A. Leineweber, E. Mittemeijer, Phase transformations in iron-nitride compound layers upon low-temperature annealing: Diffusion kinetics of nitrogen in ε- and γ'-iron nitrides. Metall. Mater. Trans. A **37**(2), 319–330 (2006)
49. S. Hosmani, R. Schacherl, E. Mittemeijer, Nitrogen absorption by Fe-1.04at.%Cr alloy: Uptake of excess nitrogen. J. Mater. Sci. **43**(8), 2618–2624 (2008)
50. K. Jung, R. Schacherl, E. Bischoff, E. Mittemeijer, Nitriding of ferritic Fe-Cr-Al alloys. Surf. Coat. Technol. **204**(12–13), 1942–1946 (2010)
51. E. Mittemeijer, *Fundamentals of Materials Science. The Microstructure-Property Relationship using Metals as Model Systems* (Springer, Berlin, 2010)
52. M. Nikolussi, A. Leineweber, E. Mittemeijer, Nitrogen diffusion through cementite layers. Phil. Mag. **90**(9), 1105–1122 (2010)
53. A. Leineweber, T. Gressmann, E. Mittemeijer, Simultaneous control of the nitrogen and carbon activities during nitrocarburising of iron. Surf. Coat. Technol. **206**(11–12), 2780–2791 (2012)
54. H. Selg, S. Meka, M. Kachel, R. Schacherl, T. Waldenmaier, E. Mittemeijer, Nitriding behaviour of maraging steel: experiments and modelling. J. Mater. Sci. **48**(12), 4321–4335 (2013)
55. E. Mittemeijer, M. Somers (eds.), *Thermochemical Surface Engineering of Steels* (Woodhead Publishing, Cambridge, 2015)
56. E. Mittemeijer, Nitriding of binary and ternary iron-based alloys, in *Thermochemical Surface Engineering of Steels* (Elsevier, 2015), pp. 313–340
57. P. Cavaliere, G. Zavarise, M. Perillo, Modeling of the carburizing and nitriding processes. Comput. Mater. Sci. **46**(1), 26–35 (2009)
58. P. Cavaliere, A. Perrone, A. Silvello, Multi-objective optimization of steel nitriding. Eng. Sci. Technol. Int. J. **19**(1), 292–312 (2016)
59. P. Cavaliere, A. Perrone, A. Silvello, Steel nitriding optimization through multi-objective and FEM analysis. J. Comput. Design Eng. **3**(1), 71–90 (2016)
60. L.A. Petrova, I.S. Belashova, A. Sergeeva, Regulation of phase composition of nitrided layers in iron and steel during thermo-gas cyclic nitriding, in *Proceeding of 24th IFHTSE Congress/European Conference on Heat Treatment, June 26–27, 2017* (Nice, 2017)
61. V. Murav'ev, Effect of residual stresses on distortion of nitrided parts. Metal Sci. Heat Treat. **17**(9), 796–798 (1975)
62. A. Zot'eva, O. Kotsyubinskii, Predicting distortion during nitriding. Metal Sci. Heat Treat. **19**(11), 986–989 (1977)
63. P. Depouhon, J. Sprauel, M. Mailhé, E. Mermoz, Mathematical modeling of residual stresses and distortions induced by gas nitriding of 32CrMoV13 steel. Comput. Mater. Sci. **82**, 178–190 (2014)
64. J. Tacikowski, J. Zyśk, *Method of Gas Nitriding*. Patent PL 85924 (Institute of Precision Mechanics, Poland, 1977)
65. Y. Lakhtin, Y. Kogan, Controlled nitriding processes. Metal Sci. Heat Treat. **20**(8), 667–671 (1978)
66. J. Ratajski, R. Olik, T. Suszko, J. Dobrodziej, J. Michalski, Design, control and in situ visualization of gas nitriding processes. Sensors **10**(1), 218–240 (2010)
67. J. Michalski, K. Burdyński, P. Wach, Z. Łataś, Nitrogen availability of nitriding atmosphere in controlled gas nitriding processes. Arch. Metall. Mater. **60**(2), 747–754 (2015)
68. J. Michalski, *Characteristics and Calculations Atmospheres for Controlled Gas Nitriding of Steel* (Institute of Precision Mechanics, Warsaw, 2010). (in Polish)
69. J. Michalski, K. Burdy, P. Wach, J. Tacikowski, Design of the gaseous nitriding processes with availability of nitrogen in nitriding atmosphere taken into consideration. Inżynieria Powierzchni **1**, 3–9 (2011). (in Polish)

70. J. Michalski, T. Frączek, Mathematical models of the change in composition of the inlet atmosphere for the assumed changes in the nitrogen potential. Inżynieria Powierzchni **25**(1), 1–9 (2020). (in Polish)
71. L. Małdziński, Controlled nitriding using a ZeroFlow process. Heat Treat. Progr. **8**, 53–55 (2007)
72. L. Małdziński, M. Bazel, M. Korecki, A. Miliszewski, T. Przygoński, Controlled nitriding using a ZeroFlow method, industrial experiences. Heat Treat. Progr. **8–9**, 19–22 (2009)
73. L. Małdziński, M. Bazel, M. Korecki, A. Miliszewski, T. Przygoński, Industrial experiences with controlled nitriding using a ZeroFlow method. Heat Treat. Progr. **7–8**, 19–22 (2009)
74. L. Małdziński, J. Tacikowski, ZeroFlow gas nitriding of steels, in *Thermochemical Surface Engineering of Steels*, ed. by E. Mittemeijer, M. Somers (Woodhead Publishing, Oxford, 2015), pp. 459–483
75. L. Małdziński, M. Hemsath, Reducing ammonia consumption and emissions using ZeroFlow gas nitriding, in *Proceedings of 28th Heat Treating Society Conference (Heat Treating 2015), October 20–22, 2015* (Detroit, 2015), pp. 223–228
76. J. Kowalska, L. Małdziński, ZeroFlow-new, environmentally friendly method of controlled gas nitriding used for selected car parts. IOP Conf. Seri. Mater. Sci. Eng. **148**, 012047 (2016)
77. Z. Haś, P. Kula, Nitrovac'79–new technology of thermo-chemical treatment of machine elements and tools. Inżynieria Materiałowa **5**(16), 127–132 (1983). (in Polish)
78. Z. Haś, *Method of Thermo-chemical Treatment of Machine Parts and Cutting Tools*. Patent PL 126128 (Lodz University of Technology, Poland, 1985)
79. Z. Haś, *Method of Subjecting Components of Machines, Mechanisms and Tools to Gas Nitriding*. Patent PL 178378 (Lodz University of Technology, Poland, 1995)
80. Z. Gawroński, Residual stresses in the surface layer of M2 steel after conventional and low pressure ('NITROVAC 79') nitriding processes. Surf. Coat. Technol. **124**(1), 19–24 (2000)
81. P. Kula, E. Wołowiec, R. Pietrasik, K. Dybowski, B. Januszewicz, Non-steady state approach to the vacuum nitriding for tools. Vacuum **88**, 1–7 (2013)
82. P. Kula, K. Dybowski, E. Wołowiec, R. Pietrasik, Boost-diffusion vacuum carburising-process optimisation. Vacuum **99**, 175–179 (2014)
83. P. Kula, Ł. Kaczmarek, K. Dybowski, R. Pietrasik, M. Krasowski, Activation of carbon deposit in the process of vacuum carburizing with preliminary nitriding. Vacuum **87**, 26–29 (2013)
84. E. Wołowiec, P. Kula, B. Januszewicz, M. Korecki, Mathematical modelling the low-pressure nitriding process. Appl. Mech. Mater. **421**, 377–383 (2013)
85. P. Jacquet, M. Devienne, A. Gokelaere, N. Duret, Use of a probe to control low-pressure nitriding treatments, in *Proceeding of 24th IFHTSE Congress/European Conference on Heat Treatment, June 26–27, 2017* (IFHTSE, Nice, 2017)
86. T. Frączek, J. Michalski, B. Kucharska, M. Opydo, M. Ogórek, Phase transformations in the nitrided layer during annealing under reduced pressure. Arch. Civil Mech. Eng. **21**(2), 48 (2021)
87. G. Perez, Low-pressure nitriding. Ind. Heat. **73**(9), 67–70 (2006)
88. D. Jordan, Vacuum solution nitriding of martensitic stainless steel holds promise. Adv. Mater. Process. **171**(7), 28–30 (2013)
89. J. Senatorski, J. Tacikowski, E. Roliński, S. Lampman, Tribology of nitrided and nitrocarburized steels, in *by G.* ed. by L. Friction, W. Technology (Materials Park, Totten (ASM International, 2017), pp.638–652
90. H. Larsson, J. Agren, Gas nitriding of high vanadium steels –experiments and simulations. Metall. Mater. Trans. A **35**(9), 2799–2802 (2004)
91. D. Herring, Principles of gas nitriding: The nitriding process. Part I. Ind. Heat. **4**, 49–52 (2011)
92. D. Herring, Principles of gas nitriding: The nitriding process. Part II. Ind. Heat. **5**, 53–58 (2011)
93. D. Herring, Principles of gas nitriding: The nitriding process. Part III. Ind. Heat. **9**, 61–63 (2011)
94. C.H. Goodman, *Crystal Growth-Theory and Techniques*, vol. I (Academic Press, New York, 1974)

95. J. Michalski, J. Tacikowski, P. Wach, Comparison and common characteristic of nitriding atmospheres in modern gas nitriding process. Inżynieria Materiałowa **5**, 459–462 (2005)
96. E. Wołowiec, *Computer Design of Heat Treatment Processes* (Lodz University of Technology, Lodz, 2013). (in Polish)
97. M. Michalak, Z. Gawroński, Nitriding process modernization on the example of chosen parts from aircraft engine. Arch. Metall. Mater. **54**(3), 773–782 (2009)
98. H. Grabke, Reaktionen von Ammoniak, Stickstoff und Wasserstoff an der Oberfläche von Eisen. Berichte der Bunsengesellschaft für physikalische Chemie **4**, 533–548 (1968). (in German)
99. H. Grabke, Evidence on the surface concentration of carbon on gamma iron from the kinetics of the carburization in CH4-H2. Metall. Trans. **1**(10), 2972–2975 (1970)
100. H. Grabke, Kinetics and mechanisms of the surface reactions between iron and gases during carburization and decarburization and nitriding and denitriding. Archiv für das Eisenhüttenwesen **46**(2), 75–81 (1975). (in German)
101. B. Lightfoot, D. Jack, Kinetics of nitriding with and without white layer formation, in *Proceedings of Conference on Heat Treatment '73, December 12–13, 1973, London* (The Metals Society, London, 1975), pp. 59–66
102. E. Wołowiec-Korecka, J. Michalski, B. Januszewicz, The stability of the layer nitrided in the low-pressure nitriding process (LPN). Coatings **13**, 257–268 (2023)
103. M. Somers, IFHTSE Global 21: Heat treatment and surface engineering in the twenty-first century. Part 14—Development of compound layer during nitriding and nitrocarburising; current understanding and future challenges. Int. Heat Treat. Surf. Eng. **5**(1), 7–16 (2011)

Chapter 6
Development of Thermo-chemical Treatment with Computational Techniques

Number rules the universe.

Pythagoras, 572–497 BC

6.1 Building a Computational Model

By quoting M. Amanowicz and W. Gasparski, the model is a description of objects of the real world, abstracting from the characteristics of objects considered to be irrelevant.[1] A variant of modelling is simulation, where a modelled phenomenon is induced and observed analogous to the phenomenon under study, but for some reason difficult to induce in natural conditions (Fig. 6.1) [2]. The areas of practical application of computer techniques in the modelling of heat and thermo-chemical treatment were first outlined by Y. Lakhtin et al. [3], who indicated the two main directions of applications: (1) the use of a computer to directly control the production process of heat treatment and (2) computer modelling of the processes occurring during heat treatment to predict the production results. This is in line with later works, including that of M. Berry and G. Linoff, investigating the general modelling of [4] phenomena. According to Berry and Linoff, the two main groups of modelling tasks include: (1) predictive issues and (2) issues of discovering the new knowledge. Predictive issues are issues of predicting or estimating the effects of phenomena on the basis of characteristics describing those phenomena. Depending on the accuracy of the selected descriptive features (called explanatory features or input data), a predictive model estimates the outcome values (called features being explained or output data) with some precision. Discovering the new knowledge is meant as extracting the relationships or values of features from the data that was collected for another purpose. Y. Lakhtin et al. predicted the first of these directions.

[1] Part of this chapter has been reproduced with permission from Ref. [1] Copyright (2018) Elsevier.

E. Wołowiec-Korecka, *Carburising and Nitriding of Iron Alloys*, Physical Chemistry in Action, https://doi.org/10.1007/978-3-031-59862-3_6

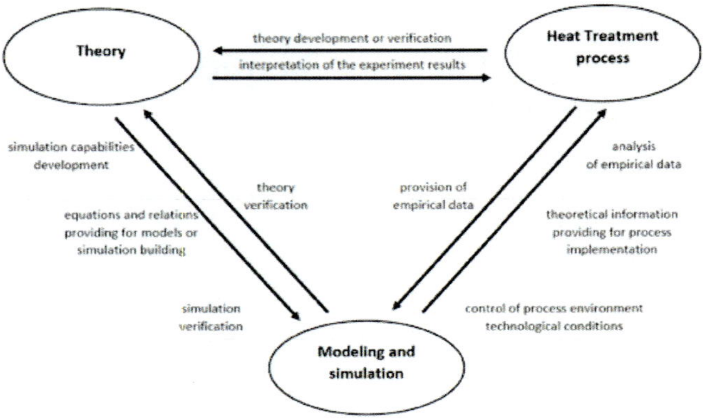

Fig. 6.1 The role of modelling and simulation in the development of the state of the art in thermo-chemical treatment

The general modelling procedure is shown in Fig. 6.2. The model-building process begins with analysing the data (selected at the data collection stage of the experiment), which usually consists of three parts: extrapolation, selection and actual analysis. The data collected in the experimentation phase usually needs to be cleaned (e.g. by removing redundant, contradictory or dubious items) and transformed into a usable form, which happens in the extrapolation phase. A subset of variables, i.e. the characteristics of the phenomenon that were considered relevant, is initially selected in the next step. The aim of these measures is to reduce the number of analysed variables to a reasonable level, allowing for effective analysis [5].

Further actions, i.e. the actual data analysis, depend on the specific problem and the research guidelines. Various models can be used to restore the nature of the considered phenomenon, ranging from simple selection of model variables with the method of linear regression to examination with graphical and statistical methods to determine the general nature of the phenomenon, its complexity and its most important characteristics [6, 7]; surely the trick is to quickly and correctly select the best one. The literature classifies modelling techniques, and thus models, into three distinct groups: conventional models (physio-mathematical), numerical models and models based on statistics (conventional or of the type *Data Mining*), each of which has its own advantages and areas of application (Fig. 6.3). The mentioned groups of methods continue to be refined and applied to the modelling of various processes and relationships, both in the production of the primary material and in further processing by materials engineering methods.

The model with the best score at the data analysis stage moves on to the final stage, where it will be applied to new data. Regardless of the choice made, the main criterion for evaluating the selection is the quality of the prediction or classification, i.e. assessing the correctness of the generated values of the modelled variables and the stability of the obtained results for different samples.

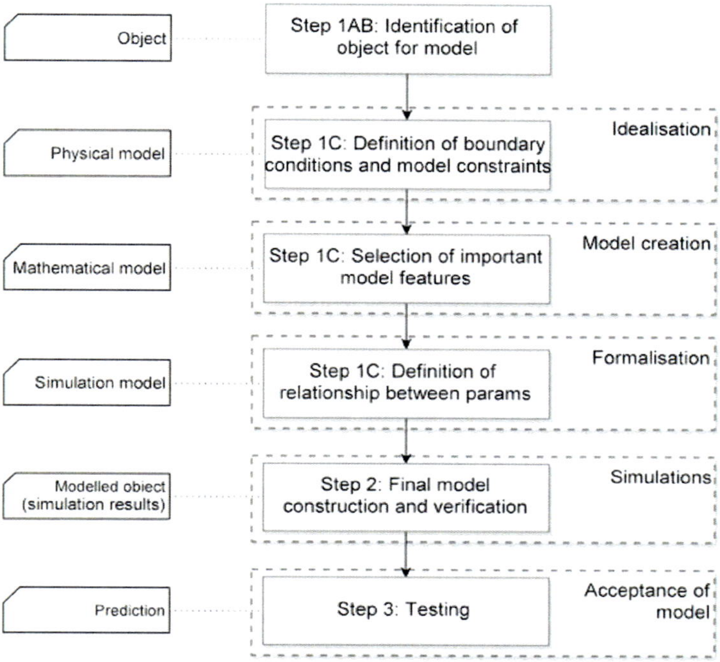

Fig. 6.2 General procedure of process modelling: (1) Preliminary analysis (A—Extrapolation, B—Preliminary choices of the variables, C—Determination of the most important features, the complexity and the nature of the phenomenon); (2) Model construction and assessment of correctness (main data analysis); (3) Applying of the model to new data. Reprinted with permission from Ref. [1] Copyright (2018) Elsevier

Conventional modelling methods

The literature most often classifies modelling methods into three distinct groups: the conventional, numerical and statistical methods. Statistical methods are divided into traditional methods and methods of discovering new knowledge (Fig. 6.3). Conventional modelling methods are usually based on fundamental laws and theories that describe the physical nature of the process.[2] Hence, the basic assumption in this type of modelling is that the physical model and the mathematical equations describing the phenomenon are perfectly known, and the equations can be solved by analytical methods. In order to use this type of modelling, a mathematical model needs to be

[2] At this point, it is important to emphasise the distinction between the terms *method* and *technique*. The method defines the general assumptions and what follows from them, while the technique delves into the technical and implementation part (e.g. methods include: observation, interview, experiment, etc. and techniques are: participant/non-participant observation, questionnaire interview, in-depth interview, telephone interview, computer-assisted interview, etc.).

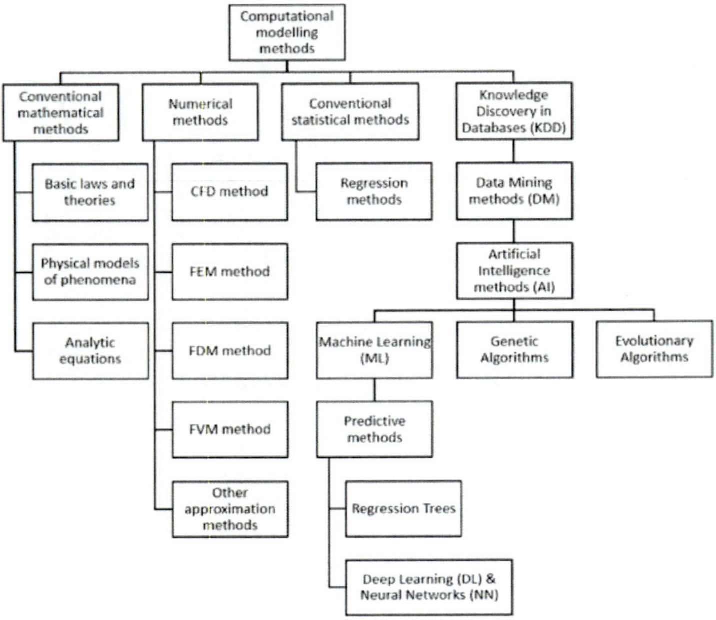

Fig. 6.3 Classification of computational methods for modelling thermal and thermo-chemical phenomena: CFD—Computational Fluid Dynamics, FEM—Finite Element Method, FDM—Finite Difference Method, FVM—Finite Volume Method, KDD—Knowledge Discovery in Databases, DM—Data Mining, AI—Artificial Intelligence, ML—Machine Learning, DL—Deep Learning

built, i.e. a description needs to be created of the properties and behaviour of the modelling objects by means of appropriate logical and formal methods employed in mathematics. When creating a mathematical model, the original is assigned to a specific computational system (most often a system of differential equations) that allows relationships to be found between the parameters appearing in the model, based on each set of values of the physical quantities considered by the model [5]. This area of modelling comprises the use of some of the conventional statistical methods, such as parametric regression. In this sense, a model is a set of rules or equations derived from the data to account for random variation, while the desire of any researcher being that the random component should tend towards zero. A simple linear model (simple linear regression) is of the form [8]:

$$y = \beta_0 + \beta_1 \cdot x + \xi \tag{6.1}$$

where β_0 and β_1 denote the structural parameters of the regression function (free expression and regression coefficient of variable y against x, respectively) and ξ denotes the random component. It is therefore not a deterministic (functional) relationship, where a change in one variable causes a well-defined change in the other variable, but a statistical relationship that does not uniquely assign values of y to given values of x, but can be accurately described using probabilistic methods. The assumptions of the linear regression model are as follows [9]:

1. The relationship between the variable being explained and the explanatory variable is linear with respect to the model parameters (β_0, β_1) (assumption concerning the form of the model).
2. The explanatory variable is not a random variable (its values are determined by the researcher) and the values of the variable are measured without error (an assumption about the variable being explained).
3. All observations have approximately similar effects on the results of the regression analysis and inference (assumptions concerning observation).
4. ξ_i have a normal distribution for all i, the average value of the random component of ξ is 0 and the variance is constant ($\text{var}(\xi_i) = \sigma^2$) for all i; ξ_i and ξ_j are independent for all $i \neq j$ (assumption concerning a random component).

As there is an analytical solution for simple linear regression, fitting the regression function to the data is done using the least squares method (LSM), where the loss function is minimised (Eq. 6.2):

$$\sum_{i=1}^{n} (y_i - \hat{y}_i)^2 = \mathbf{min} \tag{6.2}$$

where $\hat{y} = b_0 + b_1 \cdot x$ is the calculated regression model. A generalisation of simple linear regression is multifactor (multiple) linear regression with Eq. (6.3):

$$Y = \beta_0 + \beta_1 \cdot X_1 + \beta_2 \cdot X_2 + \cdots + \beta_n \cdot X_n + \xi \tag{6.3}$$

The multiple linear regression also has a strict solution achieved most easily by the least squares method. In multifactor regression, however, it is possible that explanatory variables may be related to each other, which would significantly degrade the quality of the model. The method of analytically determining the relevant variables has been described, inter alia, by L. Kukiełka [5]. An iterative way of selecting explanatory variables are stepwise methods, such as the progressive stepwise regression method described by P. Januszewski [9]: the significance of the next variable introduced is investigated with the method, and also different ordering of the variables at each step is investigated. The procedure consists of the following steps:

1. The variable is considered with the highest value of the correlation coefficient with the variable being explained until it becomes insignificant at the assumed confidence level α.
2. It is examined which variable increases the theoretical mean squared error the most, and this increase must be statistically significant. The variable that significantly increases the mean squared error is included in the model.

Table 6.1 Applications of conventional computational methods to model thermal and thermo-chemical phenomena [1, 10]

Method	Application	Ref.
Analytical equations, physical (phenomenological) models	Solution of Fick's laws	[11–15]
	Calculation of carbon activity in austenite (as a function of concentrations of alloying elements)	[16]
	Calculations of the carbon diffusion coefficient	[14, 17–20]
	Calculations of the kinetics of element diffusion in steel	[21, 22]
	Prediction of volume fractions of phase transformations	[23–26]

3. It is checked whether the explanatory variables previously introduced into the model significantly increase the mean squared error, including when it is included in the model as the second one; if not, it is rejected from the model.
4. The procedure continues until some of the rejected variables significantly increase the theoretical value of the mean squared error. The previously included variables can be discarded at each step of the calculations using the F test or the significance test of correlation coefficients.

The final model represents the optimal subset of explanatory variables assuming that the model has been correctly constructed, no potential variable has been omitted and the data have good interpretability.

The regression model is evaluated based on the coefficient of determination ($R^2 \in [0; 1]$), which indicates the percentage of the explained variation:

$$R^2 = \frac{\sum_{i=1}^{n} (\hat{y}_i - \bar{y})^2}{\sum_{i=1}^{n} (y_i - \bar{y})^2} \tag{6.4}$$

The popularisation of computers has also enabled the widespread use of a new tool for building models, i.e. logarithmic-linear analysis [27]. The primary purpose of log-linear modelling is to test the hypotheses expressed by a model whose parameters describe the distribution of variables and the relationships between the variables. The hypothesis that is formulated serves to answer specific research questions, then the expected counts or probabilities are calculated according to the model adopted. Finally, the fit of the model is compared to the observed counts or probabilities, using the G^2 statistics or Pearson's *chi-quadrat* [28]. It is a rather complex method but with great possibilities and interest in it is growing steadily [27, 29–31].

The conventional modelling of non-linear relationships, including exponential models of the type $y = \beta_0 \cdot {\beta_1}^x$ and of power models of the type $y = \beta_0 \cdot x^{\beta_1}$, is described by the extensive literature of the subject [8, 32–34].

Numerical modelling methods

Numerical methods are widely used in applied mathematics, especially where the problem under consideration does not have an analytical solution (defined by formulas) or where the application of such solutions is cumbersome due to their computational complexity (Table 6.2). The results obtained by such methods are usually approximate, but the accuracy of the calculations can be determined in advance. Numerical methods consist in determining the final result by incrementally approaching it with a sequence of approximations. As a result, the result will always contain some error, although in most cases this can be arbitrarily small. Hence, over the past three decades, numerical simulations have been broadly used in engineering problems, in particular the finite element method (FEM), due to its ability to reduce time-consuming and cumbersome trial-and-error experiments. According to D. Kim et al. [35], finite element models are now convenient for predicting the course of the carburising process, but also for calculating deformations caused by phase transformation, transformation plasticity and thermal expansion after gas quenching under pressure. Therefore, many researchers have reported that in the study of the properties of steel after thermo-chemical treatment, they have abandoned the use of analytical calculations in favour of FEM software, such as HEARTS [36], ABAQUS [37], SYSWELD [38, 39], DANTE [40], COSMAP [41] and DEFORM-HT [42–46]. However, even in commercialised computing programmes, some problems still remain without a satisfactory solution [47].

Numerical methods are extensively used in computational fluid dynamics (CFD), where they are used to solve fluid flow problems. By discretising and numerically solving the partial differential equations describing the flow, it is possible to determine approximately the distribution of velocity, pressure, temperature and other parameters in the flow. Most commercially available tools for studying computational fluid mechanics are based on the Navier-Stokes equation (the equation of conservation of mass, momentum and energy for fluid) and discretise them using the finite volume method (FVM), the finite element method (FEM) or the finite difference method (FDM).

Data mining-based methods

When knowledge of a phenomenon is incomplete, traditional data analysis methods and analytical and numerical approaches become insufficient. This is the case when a large amount of data must be analysed and a number of parameters have to be taken into account, which can have a significant impact on the features of the process or object being processed. Also in non-equilibrium processes, the relationships between process parameters and workpiece characteristics may be unknown and it is risky to make assumptions about the type of relationship that exists between them (e.g. that

Table 6.2 Comparison of selected numerical methods

	Finite element method, FEM	Finite difference method, FDM	Finite volume method, FVM	Boundary element method, BEM
Principle of operation	Applies interpolation of the function sought, for a discrete set of its nodes determined by discretising the domain of its determinacy into so-called finite elements	Applies approximation of the derivative of a function by finite differences, in a discretised space (replaces the derivatives appearing in the equations by the corresponding differential quotients)	Applies approximation of a differential equation in the form of a system of algebraic equations, for a discrete set of its nodes (a grid approximating the shape of an object); uses the Gauss-Ostrogradsky theorem to convert the integral by volume to an integral by area	Applies the fundamental solution of a differential equation
Advantages	Possibility of obtaining solutions for areas of complex shapes for which strict analytical calculations cannot be conducted; the method mainly uses symmetric matrices	Method easy to use and implement; no numerical integration	Method easy to apply for unstructured grids (e.g. in fluid mechanics)	Only the edge of the area is discretised and approximated; easy modelling of infinite and semi-infinite areas; low computational costs
Defects	A grid needs to be constructed correctly (matching the grid to area can be difficult); the domain needs to be distributed uniformly, resulting in a large number of nodes; inappropriate definition of border conditions results in loss of accuracy; not suitable for modelling infinite areas; computational process is time-consuming	A computational grid needs to be built; not suitable for modelling infinite areas; requires very fine grids; computational process is time-consuming	Mechanisms for estimating solutions to higher order functions are more difficult than FEMs	Method uses mostly non-symmetric matrices; needs existence and knowledge of solution of fundamental differential equation; difficult modelling of heterogeneous and non-linear problems

Table 6.3 Applications of numerical computational methods to model thermal and thermo-chemical phenomena [1, 10]

Method	Application	Ref.
Computational fluid dynamics fluid dynamics method, CFD	Modelling the pyrolysis of propane and acetylene under vacuum conditions	[48, 49]
	Heat flow issues	[50, 51]
	Refrigerant gas flow issues	[50, 52]
	Modelling the ammonia decomposition process	[53]
Finite element method, FEM	Temperature course during cooling	[54]
	Prediction of hardness distribution after LPC+HPGQ	[55–58]
	Prediction of phase fractions after LPC+HPGQ	[36]
	Prediction of kinetics of phase transformations during LPC+HPGQ	[59]
	Prediction of deformation due to phase transformation, transformation plasticity and thermal expansion/contraction after HPGQ	[46, 60]
	Prediction of residual stresses after HPGQ	[46, 61]
	Prediction of stresses after LPC+HPGQ	[24, 46, 61, 62]
Finite difference method, FDM	Modelling of the carburising process	[63]
	Solution of fick's laws	[64]
	Calculation of the heat transfer coefficient	[23]
Finite volume method, FVM	Modelling of HPGQ process	[51, 65]

they are linear). Such cases require a statistical approach, based on probability rather than certainty of the occurrence of a certain event.

The Knowledge Discovery in Databases (KDD) methodologies is a field of computer science encompassing tools and theories to help extract useful and previously unknown information (i.e. knowledge) from large digital data sets [66, 67]. The KDD procedure consists of several steps and one of them is data mining methods, called *Data Mining* (DM) [68]. *Data Mining* is the application of a specific algorithm to

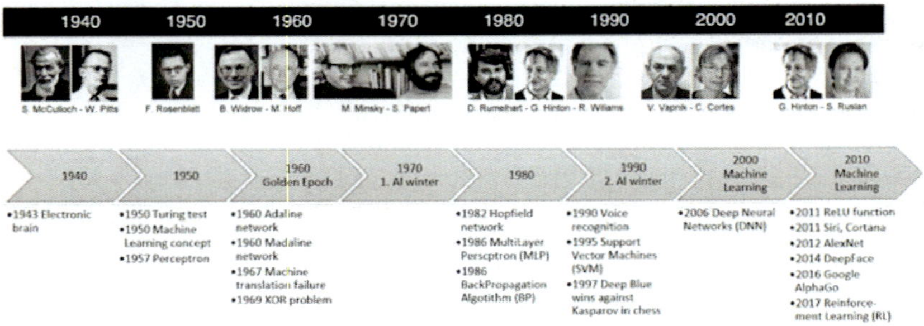

Fig. 6.4 Development of Artificial Intelligence methods. Adapted with permission from Ref. [70] Open access under a CC BY 4.0 license, https://creativecommons.org/licenses/by/4.0/

extract patterns from the data.[3] Nowadays, the use of data mining methods (DM) is a common trend in Industry 4.0 modelling and simulations. Quoting M. Berry and G. Linoff [4], *data mining* is the process of examining and analysing large amounts of data using automated and semi-automated methods with the aim of discovering meaningful patterns and rules. Model quality is evaluated with a test sample, which is usually taken from an experimental database. As mentioned earlier, DM is part of a broader set of knowledge analysis methods called knowledge discovery in databases (KDD), which include the use of data mining methods and database-related techniques. G. Migut and T. Demski list four key features that distinguish data mining from traditional statistical data analysis [69]:

- ability to analyse large data sets,
- orientation towards practical results and applications rather than the development and testing of theory,
- use of methods that are able to reproduce virtually any relationship between variables,
- prioritising the use of already existing data over performing a new experiment

A particularly fast-growing branch of *Data Mining* is artificial intelligence methods (AI); most avant-garde fields of computer science are currently evolving in this area. Although a full description of the issue is beyond the scope of this paper, it is an interesting topic to explore in more depth, as artificial intelligence methods permeate our everyday world and are readily applied even by people who do not deal with them on a daily basis. A first look at the case may cause confusion, as there are no unifying features between the particular methods. In fact, artificial intelligence is a collection of very different methods that mostly have nothing in common.

According to R. Tadeusiewicz, artificial intelligence is referred to as all the actions of machines that we would be inclined to consider as intelligent actions if they were performed by a human being [72]. The first works in the area were carried out in the 1940s (Fig. 6.4); a remote algorithm of conversation with a human being was

[3] In the literature on the topic, the terms *Knowledge Discovery Databases* (KDD) and *Data Mining* (DM) are often used interchangeably, which can be confusing to the reader.

sought that would pass the so-called Turing Test, i.e. a communication algorithm that would be indistinguishable from conversation held with another human in a remote conversation with a human via a text portal. The term Artificial Intelligence itself was not coined until 1956 by J. McCarthy or M. Minsky at a conference at Darthmounth College and the question of authorship became a source of contention in the scientific community at the time. Against this backdrop, few have noted the emergence of a new concept; the concept of machine learning developed by A. Samuel (an employee of MIT and IBM) and its first application.[4]

Machine Learning (ML) has great potential from the point of view of contemporary needs according to the Industry 4.0 concept. It belongs to heuristic methods, which means that when solving a problem, the machine assumes some initial solution, which can be drawn by the algorithm or imposed by the user. The machine applies the solution and evaluates its effect. If the effect is unsatisfactory, the machine makes some adjustment to the solution and reassesses the result. By small steps, cyclically improving its solution, the machine can arrive quite by itself at the course of action that produces the best effect and, in an age of growing computational capabilities of computers, it does not matter that it may need as many as a million iterations [72]. Specific machine learning methods include pattern recognition, decision trees, neural networks and many others. In general, they can be called learning methods.

Neural Networks (NN), of most interest to the laics, draw inspiration from the human nervous system and are widely used in non-linear modelling [73, 74]. They are now a fully-fledged modelling method and the number of reports on their applications in the literature is steadily growing [10, 75–83].[5] The correct choice of neural network architecture is particularly important in the learning process and subsequent operation [84]. Admittedly, there are no set guidelines that assign a neural network type to a specific problem, but statistical analysis of the problem indicates that some types are used more frequently for specific problems than others [1, 85, 86]. Neural networks have gone through a twofold period of oblivion and abandonment of their use (the so-called AI winter). In the early 2000s, they made a big comeback in the form of multilayer neural networks trained using the deep learning method (DL). Deep learning algorithms are capable of analysing vast amounts of unstructured data and are currently used in, for example, image retrieval on the web and natural language processing (virtual voice assistant). In 2021, the application of Deep learning in computational science and engineering was postulated by S. Saha et al. [87]. In summary, machine learning currently encompasses a whole set of predictive and data mining methods and methods based on neural networks are only a part

[4] It was a chess training programme developed between 1952 and 1962. Today, machine learning is one of the components of artificial intelligence.

[5] The history of the development of artificial intelligence sees periods of high interest in the subject, alternating with periods of quietened work. In the 1990s it was recognised that artificial neural networks had exhausted their potential for solving new problems and many researchers abandoned work on them. The re-emergence of the concept of machine learning (ML) and the rediscovery of its capabilities in processing large amounts of unstructured data (Google, IndustryFour, automatic text translators, voice processing applications, facial recognition, etc.) has resulted in a renewed explosion of works in this field.

of them (Fig. 6.3). Contemporary predictive issues include classification, regression and time series, while the discovery of new knowledge focuses on detection of variations, segmentation, cluster analysis, association and sequence analysis, collations, visualisation and text mining. The entire topic is described comprehensively in the works of S. Weiss and N. Indurkhya, B. Ratner, T. Hastie, D. Braha and others [4, 6, 7, 32, 68, 88–92].

As mentioned, predictive issues comprise situations where there is a need to develop a model that can be used to predict the value of a characteristic (dependent variable) based on a set of other characteristics (independent variables, such as process parameters). The quality of the model is assessed based on a sample set of cases for which both the independent and dependent variables are known. Predictive issues also include classification and regression. In the case of regression, apart from conventional and numerical methods, the following *Data Mining* methods are used for solving this type of problems [4, 7, 88, 89, 93]: linear regression, regression trees, linear models and generalised linear models, MARS, neural networks, boosted trees, the k-nearest neighbours method (k-NN) (which can be useful in providing missing data) and the support vector method (SVM). The following methods are used to solve classification problems: neural networks, classification trees, discriminant analysis, logistic regression, generalised linear models, general additive models, boosted classification trees, k-NN method, Naive Bayes classifier, SVM and MAR curves. A detailed description of these methods is beyond the scope of this work. These methods vary widely, making it difficult to arbitrarily choose the best one. Whether a particular method (or group of methods) is effective depends on the objectives of the analysis and the structure and quality of the data; therefore, G. Migut and T. Demski [69], while reviewing the literature on the subject, suggest that the evaluation should be aspectual, according to the aspects presented in Table 6.4.

Fuzzy Logic, alongside neural networks and evolutionary algorithms, is another modern tool for creating intelligent systems capable of generalising knowledge. Methods based on the tool work well wherever the complexity of the considered issue makes it difficult or even impossible to accurately predict the behaviour of the system. Fuzzy Controllers are used in electronic control systems (controlling train movements in subways in Japan, for fertilising crops, for directing robots), data mining tasks (investment decisions, searching databases) or building expert systems. Moreover, they allow for smooth process control, more inherent in human-controlled systems than machine-controlled ones, and thus give better results than discrete controllers [94, 95]. The essence of fuzzy systems is that they operate on fuzzy sets instead of numbers. Fuzzy sets allow to create a fuzzy model of the system, representing the relevant features, which allows to generalise of the information [96]. The fuzzing module takes the numerical values of the subsequent parameters and converts them into fuzzy values. The application module browses the base of rules, searching for those whose conditions are met. It draws conclusions from the selected rules, accepts them as true and remembers them. The sharpening module is a mirror module to the blurring module. The *blurring—inference—sharpening* scheme is repeated cyclically every time the system parameters change. The application of

Table 6.4 Review of the selected *Data Mining* methods that are helpful for solving regression and classification problems. Reprinted with permission from Ref. [69] Copyright (2012) StatSoft Poland

	Modelling methods							
	1	2	3	4	5	6	7	8
Natural support for data of different types	−	−	+	+	−	−	+	+
Lack of data	−	−	+	+	+	−	±	−
Resistance to outlier data	−	−	+	−	+	−	+	−
Scalability (N observations)	−	−	+	+	−	+	−	±
Scalability (N variables)	−	−	+	±	±	+	±	±
Resistance to variables irrelevant	−	−	+	+	−	±	−	−
Interpretability	−	−	+	±	−	±	−	+
Predictive power	+	+	−	±	+	−	+	−

1. NN—Neural Networks, 2. SVM—Support Vector Method, 3. trees, 4. MARS—Multivariate Adaptive Regression Spline, 5. k-NN—k-Nearest Neighbours method, 6. linear models (GLM, GRM, GDA), 7. boosted trees, 8. naive Bayes method

a fuzzy logic mechanism for modelling the gas nitriding process was proposed by J. Dobrodziej in his paper [97]. (Fig. 6.5).

Virtually all of the aforementioned modelling techniques have found their way into calculations supporting heat and thermo-chemical treatment (Tables 6.1, 6.3, 6.5). In recent years, the *Machine Learning* and *Predictive Maintenence* methods have been implemented to meet the expectations set by the Industry 4.0 concept. It may also be that the *Deep Learning* method, currently used for processing the natural language and the data collected by, for example, Google servers, will find its application in this area.

6.2 Models of Treatment Atmospheres

The undeniable advantage of assisted computing is that it relieves the human workload in situations where calculations are complex, lengthy or require a large number of repetitions (iterations). The last case is when the technological conditions of the process change dynamically and a tracking calculation is required, updating the instantaneous state of the process. This is the case when there is a need to monitor the composition of the treatment atmosphere and/or its potential in thermo-chemical processes. Changes in the treatment atmosphere are inextricably linked to the progression of the process.

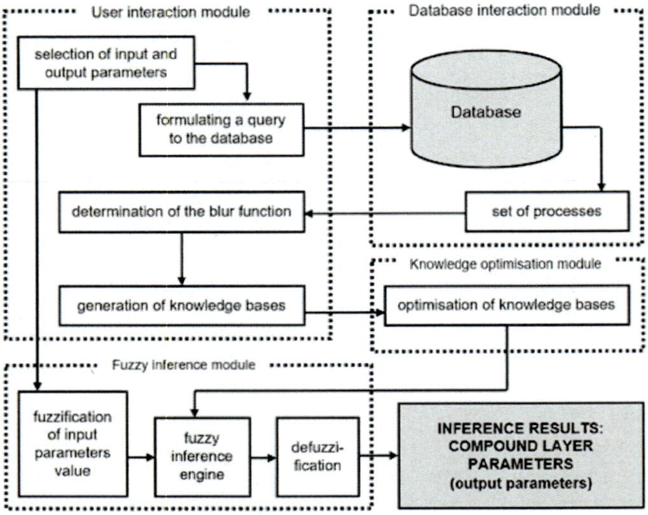

Fig. 6.5 Structure of expert system for predicting parameters of the gas-nitrided layer using fuzzy inference. Adapted with permission from Ref. [97] Open access under a CC BY 4.0 license, https://creativecommons.org/licenses/by/4.0/

Endothermic carbon-bearing atmospheres in carburising process

Carburising in gas can take place in different atmospheres. Atmospheres for endothermic carburising consist of a mixture of carburising components (CO, CH_4) and decarburising components (CO_2, H_2O), the ratio of which determines the carbon potential in the furnace. Since the driving force of carburising is determined by the gradient between the carbon potential in the atmosphere and the carbon on the steel surface, it is necessary to maintain a high carbon potential throughout the process. It is estimated that nearly 180 chemical reactions occur simultaneously[6] in an endothermic carburising atmosphere, of which only three reactions that determine the rate of carbon transfer from the carburising atmosphere to the steel surface are relevant [56, 112]:

$$2CO \rightarrow C_{(Fe_\gamma)} + CO_2 \tag{6.5}$$

$$CH_4 \rightarrow C_{(Fe_\gamma)} + 2H_2 \tag{6.6}$$

[6] Acc. to A. Burghard and T. Zaleski, in a closed system containing <J> reactants formed from <N> elements or groups of elements stable under these conditions, the most R = J − N independent reactions can proceed [111].

Table 6.5 Applications of knowledge discovery methods (KDD), including data mining methods (DM) for modelling heat and thermo-chemical phenomena [1, 10]

Method	Application	Ref.
Traditional statistical methods		
Analysis of variances	Assessment of the significance of factors influencing the phenomenon	[6, 7]
Regression methods	Determination of the diffusion coefficient of carbon	[14, 17–20, 98]
Data mining methods		
Regression neural networks	Prediction of multiphase diffusion during LPN process	[99]
	Predicting the hardness of the nitrided layer after LPN process	[99]
	Predicting the creation and dissolution of carbide precipitates during LPC process	[100–102]
	Predicting the distribution of hardness after HPGQ	[103]
	Optimisation of heat and thermo-chemical treatment process parameters	[102, 104, 105]
	Modelling the hardenability and other properties of steels	[106–108]
	Determination of diagrams of TTT phase transformations	[77]
Classificational neural networks	Categorisation of materials according to their characteristics	[99]
Genetic algorithms	Optimisation of heat treatment parameters	[104]
	Design of alloy materials	[109]
Evolutionary algorithms	Optimisation of heat treatment process parameters	[110]

$$CO + H_2 \leftrightarrow C_{(Fe_\gamma)} + H_2O \tag{6.7}$$

The carbon potential (C_P) in the atmosphere of a gas carburising furnace is defined as the amount of carbon in the furnace atmosphere that is in thermodynamic equilibrium with the carbon content at the steel surface. The carbon potential of the atmosphere is calculated from the chemical reactions (6.5), (6.6) and (6.7) for a given carburising temperature and gas chemistry. As reported by O. Rowan, under carburising conditions, the carbon activity (a_C) and the reaction speed factor (k)

for these reactions are described by Eqs. (6.8), (6.9) and (6.10), while the effective activity of the carbon and its potential in the furnace atmosphere is determined by the reaction with the highest reaction speed [112].

$$a_{C_1} = \frac{P_{CO}^2}{P_{CO_2}} \cdot \exp\left(\frac{20530.65}{T} - 20.98\right), k_1 = 184 \cdot P_{CO_2} \cdot \left(\frac{P_{CO_2}}{P_{CO}}\right)^{-0.3} \cdot \exp\left(\frac{-22400}{T}\right) \quad (6.8)$$

$$a_{C_2} = \frac{P_{CH_4}}{P_{H_2}} \cdot \exp\left(\frac{10949.68}{T} - 13.31\right), k_2 = 1.96 \cdot 10^{-2} \cdot P_{H_2}^{1.5} \cdot \exp\left(\frac{-17600}{T}\right) \quad (6.9)$$

$$a_{C_3} = \frac{P_{CO} \cdot P_{H_2}}{P_{H_2O}} \cdot \exp\left(\frac{16333.11}{T} - 17.26\right), k_3 = \frac{4.75 \cdot 10^4 \cdot \exp\left(\frac{-27150}{T}\right) \cdot \frac{P_{H_2O}}{\sqrt{P_{H_2}}}}{1 + 5.6 \cdot 10^6 \cdot \exp\left(\frac{-12900}{T}\right) \cdot \frac{P_{H_2O}}{\sqrt{P_{H_2}}}} \quad (6.10)$$

O. Rowan maintains that for a typical composition of an endothermic atmosphere (19.8% CO, 0.1% CO_2, 40% H_2, 0.3% H_2O, the rest N_2) at the temperature of 925°C, the carburising reaction (6.7) is several orders of magnitude faster than the reactions of (6.5) and (6.6), hence the reaction (6.7) determines the rate of carbon adsorption during the process. The effective carbon activity calculated from Eq. (6.10) is then related to the carbon potential in the atmosphere using the model described by M. Bannister and S. Ban-Ya et al. [113, 114]:

$$\log_{10} a_C = \frac{3770}{T} + 2.72 \cdot \log_{10} - 10.525 + \frac{3860 \cdot w}{T} + \log_{10}\left(\frac{w}{1-w}\right) \quad (6.11)$$

$$C_P = \frac{1201 \cdot w}{55.58 + 12.01 \cdot w} \quad (6.12)$$

where a_C—is the carbon activity in austenite, T—the temperature [K], w—percentage mass of carbon in austenite at the steel surface.

A method of modelling the mass exchange issues during gas carburising can be found in the work of the team of R. Sisson and O. Rowan [112, 115–119]. The modelling of the atmosphere of carburising reaction and carbon potential calculations, and the modelling of the endothermic thermodynamics of the gas carburising atmosphere with a conventional mathematical apparatus, are presented in the work [115]. The modelling of the mass exchange kinetics at the gas-steel boundary and the modelling of carbon diffusion in steel is presented in the work of [116, 117, 119]. The models developed were used to predetermine (predict) the composition of atmospheric gases at the stage of carburising enrichment, the kinetics of carbon transfer at the gas-steel boundary and the diffusivity of carbon in austenite for different process conditions and alloy compositions (see Subsect. 6.3).

Carbon-bearing atmospheres based on aliphatic hydrocarbons

Unlike atmospheric carburising (in an endothermic atmosphere), low-pressure carburising takes place with excess carbon in the atmosphere. The carbon used to saturate the diffusion austenite is obtained in the process of thermal decomposition of aliphatic hydrocarbons, often propane (C_3H_8), whose decomposition at normal pressures and high temperatures occurs avalanchingly through intermediate stages [120]. In order to control the rate of reaction of the decomposition of this hydrocarbon, the dilution of propane with nitrogen or the lowering of the total pressure to the level of the so-called low vacuum is used. More recent solutions employ unsaturated aliphatic (short-chain) hydrocarbons-acetylene and ethylene (C_2H_2, C_2H_4) and mixtures thereof—instead of propane [121–123].

Kinetic models for the decomposition of light hydrocarbons used in low-pressure carburising, such as acetylene, ethylene and propane, have been presented in numerous papers by R. Khan, A. Matsugi, K. Yada and other researchers [48, 49, 52, 124–131], using the numerical methods (CFD) available in DETCHEM [132], HOMREA and Ansys FLUENT [65]. A comparison of these reactive flow modelling tools was presented in R. Khan's work [124].

Another tool for investigating the thermo-chemical properties of molecules and free radicals in the gas phase is THERGAS [131], used, for example, by T. Bensabath et al. to model the kinetic formation of toxic polycyclic aromatic hydrocarbons (PAHs) from pyrolysis under the conditions of low-pressure gas carburising [127].

Ammonia-based nitrogen-bearing atmospheres

Nitrogen in the nitriding process is produced by the decomposition of ammonia. The decomposition of ammonia is a reaction taking place in the gas phase and the expression of the reaction rate can usually be expressed in terms of the partial pressures of the reacting components. The literature on the subject contains many reports devoted to the chemical kinetics of the ammonia decomposition reaction, most often stating that the reaction rate primarily depends on the partial pressure of ammonia [133–136]. S. Yin et al. have presented a concise overview of the kinetics of the reaction and of the mechanism of ammonia decomposition, but there are still very few predictive modelling attempts that have been published to elucidate the decomposition chemistry [136, 137]. K. Alhumaizi et al. have analysed mathematically the one-dimensional reaction-diffusion-convection system using standard reductive computational techniques (FDM, FEM) [138]. In 2005, A. Waghode et al. proposed a model for the kinetics of the ammonia decomposition reaction, in which the flow model (flow equations) was based on the finite element method (FEM) [137]. I. Ozdemir and F. Akar utilised the simulations of computational fluid dynamics (CFD) to study the effect of flow direction on the rate of ammonia decomposition in the nitriding process [139].

The type and composition of the nitriding atmosphere, in addition to the temperature, has a dominant influence on the kinetics of the gas nitriding process and, consequently, on achieving the desired physiochemical and operational properties of the nitrided layers (thickness and phase composition of the layers). Precise control of the nitriding atmosphere in the nitriding process is greatly facilitated when it is possible to predict the nitrogen potential (N_P), the rate of ammonia dissociation (ω) and the atmosphere composition and output at the inlet to the furnace chamber. The conditions for calculations become complex if the nitrogen atmosphere consists of more than one component, such as a three-component atmosphere containing nitrogen, ammonia and dissociated ammonia.[7] Assumptions for solving this problem were presented by J. Michalski et al., giving a method for estimating the characteristics of nitrogen potential (N_P), atmosphere output and volume fractions of ammonia, nitrogen and dissociated ammonia [141]. It is particularly important to characterise the nitrogen potential, because the way it changes as a function of duration and process temperature determines the structure and growth kinetics of the nitrided layer. For a temperature range of 350–590°C, the atmosphere $a_0NH_3/b_0N_2/c_0NH_{3diss}$ diluted with an atmosphere in the form of $a_xNH_3/b_xN_2/c_xNH_{3diss}$ to the final form $a_1NH_3/b_1N_2/c_1NH_{3diss}$ was modelled as follows [141, 142]:
(a) equilibrium potentials as a function of temperature:

$$N_P^{\alpha/\gamma'}(T) = \exp\left(\frac{4556}{T} - 12.88\right)[\mathrm{p}^{-0.5}] \tag{6.13}$$

$$N_P^{\gamma'/\varepsilon}(T) = \exp\left(\left(\frac{60536}{T} - 56.85\right)^{0.5} - 9.63\right)[\mathrm{p}^{-0.5}] \tag{6.14}$$

(b) nitrogen potential as a function of ammonia dissociation (ω), the content of ammonia (a), nitrogen (b) and dissociated ammonia (c) in a three-component inlet atmosphere with the composition $a\mathrm{NH_3}/b\mathrm{N_2}/c\mathrm{NH_3}$:

$$\begin{aligned} N_P(\omega, a, b, c) &= \\ = \frac{(0.75\cdot\omega - 0.375\cdot c\cdot\omega + 0.75\cdot c)\cdot(1+a) - 1.5\cdot(1-b-c) - 0.75\cdot c}{(0.75\cdot c - 1.5)\cdot(0.75\cdot\omega - 0.375\cdot c\cdot\omega + 0.75\cdot c)^{1.5}} \\ a+b+c = 1;\ 0 < \omega \le 1;\ \omega > 0 \end{aligned} \tag{6.15}$$

where ω—the rate of dissociation of ammonia in the nitriding atmosphere, expressed as the ratio of the doubled volume of the part of ammonia s in the inlet atmosphere that dissociated during the nitriding process to the volume of the outlet atmosphere $\omega = 2s/(1+s)$. In the general case, a, b, c and ω are a function of time,

[7] This type of atmosphere combines the properties of two-component atmospheres NH_3, N_2, and can be used as a reference for atmospheres containing only ammonia [140].

(c) changes in the volume fractions of ammonia, nitrogen and dissociated ammonia as a function of temperature, furnace retort volume, atmosphere output and duration:

$$a(F,t) = a_0 \cdot \exp\left(\frac{-e \cdot F \cdot t}{V_P}\right) + a_1 \cdot \left(1 - \exp\left(\frac{-e \cdot F \cdot t}{V_P}\right)\right) \tag{6.16}$$

$$b(F,t) = b_0 \cdot \exp\left(\frac{-e \cdot F \cdot t}{V_P}\right) + b_1 \cdot \left(1 - \exp\left(\frac{-e \cdot F \cdot t}{V_P}\right)\right) \tag{6.17}$$

$$c(F,t) = c_0 \cdot \exp\left(\frac{-e \cdot F \cdot t}{V_P}\right) + c_1 \cdot \left(1 - \exp\left(\frac{-e \cdot F \cdot t}{V_P}\right)\right) \tag{6.18}$$

where $a(F,t)$, $b(F,t)$, $c(F,t)$—respectively, volume fractions of ammonia, nitrogen, dissociated ammonia after duration t, a_0, b_0, c_0—respectively, volume fractions of ammonia, nitrogen, dissociated ammonia in the output atmosphere, a_1, b_1, c_1—respectively, volume fractions of ammonia, nitrogen, dissociated ammonia in the dilution atmosphere, Q—total flow rate of the dilution atmosphere, V_P—volume of the furnace chamber, t—nitriding duration, e—constant equal to 2.71828,
(d) the duration for the components of the ternary atmosphere to reach the target composition:

$$t_a = -\frac{V_P}{F} \cdot \ln\left(\frac{a_1 - a_x}{a_0 - a_x}\right) \tag{6.19}$$

$$t_b = -\frac{V_P}{F} \cdot \ln\left(\frac{b_1 - b_x}{b_0 - b_x}\right) \tag{6.20}$$

$$t_c = -\frac{V_P}{F} \cdot \ln\left(\frac{c_1 - c_x}{c_0 - c_x}\right) \tag{6.21}$$

Most of the model was built using conventional analytical equations. An evolutionary algorithm was used to optimise the changes in the composition of the dilution atmosphere (coefficients a_x, b_x, c_x), according to the method described in D. Goldberg's work [143].

Models of ammonia and acetylene decomposition and the accompanying pyrolysis and formation of soot particles under high temperature and reduced pressure conditions were investigated by E. Vyazmina et al. using numerical models available with the CHEMKIN tool [144, 145].

6.3 Models of Single- and Multiphase Diffusion

Single-phase diffusion

The phenomena of single-phase diffusion in thermo-chemical treatment are particularly evident in the process of diffusion saturation of the surface layer of steel with carbon (carburising). Carburising consists in heating low-carbon steel to a temperature above A_3 (the temperature at which homogeneous austenite exists), annealing it in a suitable carburising medium for the time required for carbon diffusion to occur in the material to achieve the required effective layer thickness (CD_{Eff}), and then cooling. A surface layer with a favourable distribution of carbon concentrations is produced as a consequence.

For atmospheric gas carburising, the effect of time on the depth of the layer thickness is interdependent on the carburising temperature and is often estimated using F. Harris' equation [146]. However, this equation assumes the existence of saturated austenite on the surface of steel, and when the carbon content on the surface is less than the saturation limit, the resulting layer thickness is less than assumed. In fact, the efficiency of the gas carburising process depends on the effective control of three main variables: carburising atmosphere, temperature and process duration. The literature on the topic contains an extensive description of the influence of these parameters on the kinetic factors of carburising, i.e. mass transfer coefficient and carbon diffusion coefficient in steel [14, 147–156].

The mass transfer coefficient (of carbon) in gas carburising is a function of the chemical composition of the carburising atmosphere, the carburising potential, the temperature and the carbon content at the surface [148–153]. It is accompanied by the carbon diffusion coefficient in austenite, also determining the rate of carbon transport, which is strongly influenced by the carburising temperature and the concentration of carbon in the steel [14, 147, 154]. Although the mechanism of mass transport in carburising appears to be known and understood, the results of carburising at the industrial scale reveal that the profiles of carbon concentrations and the depths of carburising in many cases deviate from the predicted ones, and it is only with the support of computer computations that improvements in this state become feasible [112].

Low-pressure carburising of steel differs from endothermic carburising in the phenomena taking place during the transport of carbon atoms from the atmosphere to the steel surface. It takes place under conditions deviating from thermodynamic equilibrium. The reduced pressure of the carbon-bearing hydrocarbons is also important, significantly reducing the generation of deposition on the steel surface in the form of carbonaceous compounds. However, residual surface phenomena analogous to atmospheric carburising still occur [157].

In summary, irrespective of pressure, the development of carbon profiles in austenite depends on both the phenomena occurring at the gas-metal boundary and the diffusion of carbon in austenite. The phenomena at the gas—steel boundary, amongst which the catalytic interaction between the carburising atmosphere and the surface, implying a corresponding rate of change in the surface carbon concentration, plays a major role, are represented in the mathematical model as a flux (J_C) of atomic carbon across this boundary:

$$J_C = -\beta_C \cdot \left[C_A - (C(t))_S\right] \tag{6.22}$$

where J_C—flux of carbon atoms through the surface of the carburised material, C_A—carbon concentration, equilibrium with the carburising atmosphere, $(C(t))_S$—current concentration of carbon at the surface of the material, β_C—carbon transfer coefficient across the gas—metal interphase boundary. Carbon transport in austenite is described based on Fick's diffusion laws, the first of which determines the flux of the diffusing component:

$$J_C = -D_C \cdot \nabla f(C) \tag{6.23}$$

where D_C—diffusion coefficient of carbon in austenite, the value of which depends on temperature, carbon concentration and content of alloying additives, $\nabla f(C)$—carbon concentration gradient in austenite, C—carbon concentration in austenite.

The second law describes a change in carbon concentration as a function of time:

$$\frac{\partial C}{\partial t} = \mathrm{div}(\nabla I_C) \tag{6.24}$$

The above laws for unidirectional diffusion come down to the following form:

$$J_C = D_C \frac{\partial C}{\partial x}, \tag{6.25}$$

and

$$\frac{\partial C}{\partial t} = \frac{\partial}{\partial x}\left(D_C \frac{\partial C}{\partial x}\right). \tag{6.26}$$

The influence of surface phenomena on the diffusion profile of carbon in austenite is taken into account by solving Fick's 2nd law with the border condition resulting from the comparison of the formulae (6.23) and (6.25):

$$-D_C \left.\frac{\partial C}{\partial x}\right|_{x=0} = \beta\left[C_A - (C(t))_S\right], \tag{6.27}$$

Thus, the problem of finding the carbon concentration profiles in austenite as a function of time in the carburising process reduces to solving Eq. (6.26) with a border condition (6.27) and an initial condition (6.28):

$$[C(x,\ t = 0)] = 0. \tag{6.28}$$

The border condition for the low-pressure carburising process (Eq. 6.27) is different from that for the atmospheric carburising process. While in the process of atmospheric carburising the carbon potential can be monitored and controlled, during low-pressure carburising process, the carbon concentration at the surface increases rapidly and can reach maximum solubility. The carbon concentration only decreases with carbon diffusion. P. Kula et al. proved that a long stage of saturation during the process not only causes precipitation phenomena (carbides across grain boundaries), but also deposition (carbonaceous precipitate) on the surface of components [11, 157]. R. Gorockiewicz, A. Łapiński and others have confirmed these observations [158–160].

The constant value of the mass transport coefficient is applicable in most cases of carburising (atmospheric and low-pressure carburising) because, as the carbon potential approaches the solubility limit of carbon in austenite, with a carbon content greater than 0.5%, the value of the coefficient β_C becomes dependent on temperature and has little relation to the composition of the carburising gas [161, 162]. However, the hydrocarbon gases added to the furnace chamber during the saturation stage create a very high carbon potential and carbonaceous substances may be created as a result. For this reason, the border condition of mass balance is not applicable to the carburising process. However, a relationship describing a constant flux through the surface can be used. The amount of carbon produced by surface reactions is assumed to be equal to the mass flow rate for diffusion, described by Fick's first law [161]:

$$J_C|_{x=0} = -D_C|_{x=0} \cdot \left.\frac{\partial C}{\partial x}\right|_{x=0} \tag{6.29}$$

where $x = 0$ at the gas—steel boundary.

In summary, the problem of diffusion in the solid state can be considered on the basis of the general phenomenological laws of physical chemistry and by solving the corresponding equations illustrating these laws in relation to time and space [64, 163–165]. The thickness of the carburised layer and the distribution of carbon in the diffusion zone depend on the process parameters and mass transport. The latter is described by two main parameters: the mass transfer coefficient across the gas-solid boundary $\beta_C(t)$ and the carbon diffusion coefficient D_C, which depends on the alloying elements. As it is needed to iteratively solve the diffusion equations multiple times as the process proceeds, control of the phenomenon is only possible with computational support, i.e. by building a mathematical model, with a system of equations solved numerically. Therefore, for the control of the vacuum carburising process, it is necessary to determine the effective diffusion coefficient, which is developed experimentally [166, 167].

The first models of carburisation kinetics were based on a one-dimensional form of the law of conservation of mass (Fick's second law) and the concentration-dependent diffusivity of carbon. A general analytical model of low-pressure carburising based on the Devian-Osterend equation was presented by S. Tsepov [168] in 1979. K. Bongartz et al. used the finite difference method [63, 169, 170]. More complex analytical models for calculating the effect of carburising on steel properties were described by J. Morral et al. [21, 22, 171], P. Cavaliere et al. [172] and M. Semenov et al. [13, 16]. Currently, the solution to the problem in most of the studies described is to calculate a function $c(x, t)$, i.e. a function of carbon concentration varying with time and distance from the surface. Ł. Kołodziejczyk has described a model in which the diffusion coefficient of carbon D_C in austenite depends on temperature, carbon concentration and the presence of alloying additives [173]. The activity of carbon in austenite in the presence of alloying additives influences the carbon diffusion coefficient and allows to calculate the maximum solubility of carbon in austenite (C_{Max}) at temperature T for a given material. The equations of the model can be solved using an analytical method (as in the work of AWT [174]).

The effect of process parameters, steel composition and initial surface roughness on the kinetics of mass exchange during carburising were modelled in the works of O. Rowan et al. [118, 119, 175]. The works of Y. Wei and L. Zhang complement that of O. Rowan because, while O. Rowan's work mainly uses mathematical-physical models, Y. Wei et al. and L. Zhang consider the progress in gas carburising (and low-pressure carburising, for the gases C_2H_2 and C_3H_8) of diffusion by the numerical finite difference method (FDM) [161, 176]. A numerical model of vacuum carburising with acetylene was presented by S. Wei et al. [177].

Functions describing the relationship between the diffusion coefficient of carbon D_C and temperature T or duration t have been presented in many articles [14, 17–20, 178–184]. All researchers agree that the correct relationship is crucial for the model to be consistent with actual experiments, although many papers present different relationships. The first studies of such a model were based on the carbon diffusion coefficient formula proposed by A. Wells et al. [17]. This relationship was found to be insufficient, so their own formulas were proposed by G. Leyens [19], J. Goldstein [14] and G. Tibbets [20] (see Table 6.6). More works related to this issue are discussed by A. Ochsner et al. in [12] and O. Rowan [112].

A working model of gas carburising (*CarbTool*) was described by L. Zhang and Y. Wei et al. [161, 176], and other commercially available models of atmospheric carburising were presented by D. Herring in his paper "*Simulation software. Part one: atmosphere carburizing*" [186]. Commercialised models of thermal-diffusion processes carried out under reduced pressure are also described in the works of O. Rowan, M. Jaster, P. Kula and others (Table 6.7).

Low-pressure carburising is an example of scientific research whose progress is inseparably linked to modelling and numerical calculation. It was only possible to map the kinetics of the process running at reduced pressure using a numerical calculation method. Computational support is of particular importance in non-equilibrium processes, where transient states of the process flow are present. In this case, the primary objective of the simulation is to predict the process flow and final product properties, thereby ensuring repeatable process results. As a result, there has been growing interest in recent years in mathematical descriptions of heat and thermo-chemical treatment processes, both theoretically and experimentally derived.

Work on low-pressure carburising was undertaken in 2001 by a research consortium of the Lodz University of Technology and Seco/Warwick [173, 197–199].

Table 6.6 Empirical equations of the carbon diffusion coefficient in austenite. Reprinted with permission from Ref. [1] Copyright (2018) Elsevier

Ref.	Carbon diffusivity in austenite	
[17]	$D_C(T,C) = (0.47263 - 0.55169 \cdot C + 0.22848 \cdot C^2 - 0.02965 \cdot C^3) \cdot \exp(-\frac{153593-15370 \cdot C-10727 \cdot C^2+2432.9 \cdot C^3}{R_J \cdot T})$	(6.30)
[18]	$D_C(T,C) = 0.5 \cdot \exp(-30x_C) \cdot \exp(-\frac{38300-1.9 \cdot 10^5 x_C+5.5 \cdot 10^5 x_C^2}{R_C \cdot T})$	(6.31)
[178]	$D_C(T,C) = 0.668 \cdot \exp\left(-\frac{37460}{R_C \cdot T}\right)$	(6.32)
[179]	$D_C(T,C) = 0.49 \cdot \exp\left(-\frac{36000}{R_C \cdot T}\right)$	(6.33)
[179]	$D_C(T,C) = (0.04 + 0.08 \cdot C) \cdot \exp\left(-\frac{31350}{R_C \cdot T}\right)$	(6.34)
[20]	$D_C(T,C) = 0.47 \cdot \exp(-1.6 \cdot C) \cdot \exp\left(-\frac{37000-6600 \cdot C}{R_C \cdot T}\right)$	(6.35)
[180]	$D_C(T,C) = 0.4718 \cdot \exp\left(-\frac{180590}{R_C \cdot T}\right)$	(6.36)
[180]	$D_C(T,C) = 0.0052 \cdot \exp\left(-\frac{127950}{R_C \cdot T}\right)$	(6.37)
[181]	$D_C(T,C) = 0.0077 \cdot \exp\left(-\frac{107400}{R_C \cdot T}\right)$	(6.38)
[182]	$D_C(T,C) = 0.02 \cdot \exp\left(-\frac{112000}{R_C \cdot T}\right)$	(6.39)
[184]	$D_C(T,C) = (0.146 - 0.036 \cdot C \cdot (1 - 1.075 \cdot Cr) - 0.0315 \cdot Mn + 0.0509 \cdot Si - 0.0085 \cdot Ni + 0.3031 \cdot Mo - 0.052 \cdot Al \cdot \exp(-\frac{144300-15000 \cdot C+370 \cdot C^2-4366.3Mn+4050.7 \cdot Si}{R_J \cdot T} \frac{-1240.7 \cdot Ni+7726 \cdot Cr+12126.6 \cdot Mo-6788.6 \cdot Al}{R_J \cdot T})$	(6.40)
[14]	$D_C(T,C) = (0.07 + 0.06 \cdot C) \cdot \exp\left(-\frac{32000}{R_C \cdot T}\right)$	(6.41)
[20]	$D_C(T,C) = 0.38 \cdot \exp\left(-\frac{-37000-6600 \cdot C}{R_C \cdot T}\right)$	(6.42)
[19]	$D_C(T,C) = 0.68 \cdot e^{-\left(\frac{18900}{T}+0.38\right)+\left(\frac{4300}{T}-2.63\right) \cdot C^{1.5}}$	(6.43)
[183]	$D_C(T,C) = (1.876 \cdot 10^{-11} \cdot T^2 - 4.468 \cdot 10^{-8} \cdot T + 2.669 \cdot 10^{-5}) \cdot \exp((-0.0034 \cdot T + 5.06) \cdot C)$	(6.44)
[185]	$D_C(T,C) = \frac{1.43 \cdot \exp\left(-\frac{19700}{T}+0.002242 \cdot \exp\left(\frac{6790}{T}\right) \cdot C\right)}{1-0.232 \cdot C}$	(6.45)

where D_C—carbon diffusion coefficient $\left(\frac{cm^2}{s}\right)$, C—carbon concentration, x_C—carbon mole fraction, R_C—gas constant 1.986 $\frac{cal}{mol \cdot K}$, R_J—gas constant 8.314 $\frac{cal}{mol \cdot K}$, T—temperature (K)

The results of the work were patented under the name *FineCarb* and complemented with the computational tool *SimVac* [123, 200, 201]. The mathematical and physical model of vacuum carburising was based on six assumptions given by the authors of the papers *Application of a numerical model in the design of vacuum carburising* and *Mathematical modelling of the vacuum carburising process* [173, 202]:

1. The carburising atmosphere during the stage of saturation has unlimited carbon available for the process.

Table 6.7 Commercially available software for design and simulation of thermo-chemical treatment. Reused with permission from Ref. [1] Copyright (2018) Elsevier

Products	Supplier	Specifications	Ref.
Carb-o-Prof	Ipsen, USA	Atmospheric gas curburising (*)	[187]
CarbCalcII	Super Systems, USA	Atmospheric gas curburising (*)	[188]
VC-Sim	ALD Vacuum Technologies, Germany	Low-pressure curburising (*)	[189]
CarbTool	Worcester Polytechnic Institute (CHTE), USA	Atmospheric gas curburising and low-pressure curburising	[190, 191]
SimVac	Seco/Warwick & Lodz University of Technology, Poland	Low-pressure curburising (*)	[192, 193]
Infracarb	ECM Technologies, France	Low-pressure curburising (*)	[194]
AvaC	Ipsen, USA	Low-pressure curburising (*)	[195]
SimLPN	Lodz University of Technology, Poland	Low-pressure nitriding	[196]

*–industrially used software

2. The composition of the atmosphere remains constant throughout the process.
3. Carbon atoms are released by a catalytic interaction between the atmosphere and the carburised surface.
4. Carbon is transported into the material as a result of diffusion according to the Fick equations.
5. The diffusion coefficient of carbon in austenite depends on temperature, carbon concentration and the presence of alloying additives.
6. A semi-infinite area is considered in calculations.

Currently, the computer numerical tool *SimVac* (Figs. 6.6 and 6.7) is an integral part of the vacuum carburising technology *FineCarb* [55, 203]. The software is an advanced simulation tool that enables to design vacuum carburising and high-pressure oil and gas quenching processes, as well as to analyse and optimise these processes without the need for time-consuming and costly test processes [55, 204]. Depending on the user's needs, it presents the result of implementing a preset process in the form of a carbon profile and a layer hardness profile, or it designs the process itself according to the preset requirements of the layer.

As the amount of experimental data grows every year (more and more stages of manufacturing are equipped with sensors collecting data on parameters and end results of processes and the cost of archiving, storing and processing data is declining), methods for statistical processing of large amounts of data are gaining practical

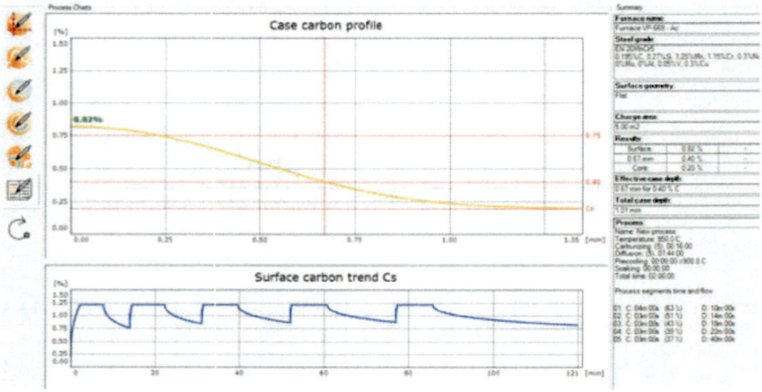

Fig. 6.6 Simulator of the kinetics of carburised layer growth

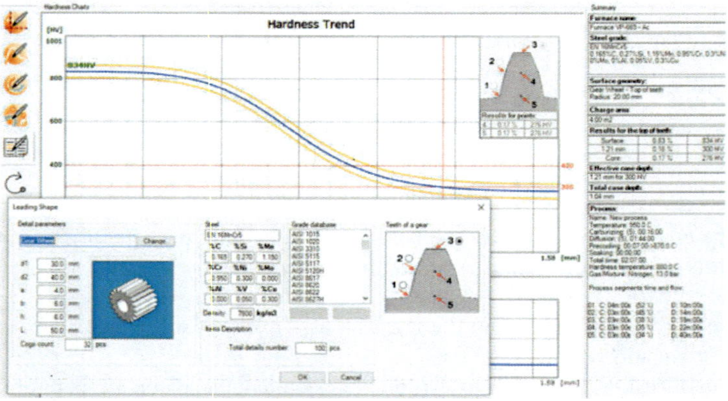

Fig. 6.7 Parameters of toothed element in simulation of high-pressure gas quenching process after low-pressure carburising process

importance. A method of analysing the extensive experimental data (*Big data*) of studies of high-temperature vacuum carburising phenomena using statistical methods (genetic algorithms) was presented by S. Hiramatsu et al. (Toyota Motors Corp.) [205]. The model of predictive carburising presented here worked with an accuracy of 10%. A model of precipitation phenomena in non-equilibrium low-pressure carburising conditions was described by E. Wołowiec et al. using the method of artificial neural networks [83, 100]. A similar method was used by K. Genel et al. to model the properties of borided layers in AISI W1 steel [206]. Neural networks are also an effective tool for solving optimisation problems [101, 207].

Multiphase diffusion

From the point of view of this work, the multiphase diffusion occurring during nitriding processes is of interest. This is a commonly used treatment, as mentioned in Chaps. 2 and 5. Considering the gas varieties conducted in conditions of atmospheric pressure, controlled gas nitriding by the NITREG method [208] and, in recent years, controlled nitriding by the ZeroFlow method [209–211] are commonly used. There has been growing interest for several years in nitriding under lowered pressure (LPN) [86, 99, 196]. As far as industrial practice is concerned, manufacturers of equipment for all these treatments are following global trends by computerising process control methods. This element influences the optimisation of conducting the process in terms of time, media consumption and, above all, treatment results.

The most important parameters influencing the kinetics of nitrided layer growth and its phase composition in gas nitriding include: nitrogen potential, temperature, composition and flow rate of the nitriding atmosphere; composition of the nitrided material; and surface preparation for nitriding. A nitrogen concentration gradient, which is the driving force behind the diffusion of nitrogen in iron Fe_{α}, is created as a result of transferring the mass of nitrogen from the atmosphere to the steel surface. The diffusivity of nitrogen in the nitriding process was examined by Y. Lakhtin, showing the differences between the values of diffusion coefficients for different nitrogen phases [212, 213]. The growth rate of each phase is determined by the intensity of the diffusion processes in the phases existing in the layers below and above. By analysing the diffusion coefficients D_N, D_0 and Q for all phases of the nitrided layer, it was revealed that nitrogen diffusion occurs most easily in the α phase and is strongly hindered in the γ' phase. For equal values of the D_0 coefficients, the diffusion coefficient of nitrogen in the α phase (D_N^{α}) should exceed the $D_N^{\gamma'}$ and D_N^{ε} coefficients many times, because the activation energy of Q in the α phase is twice as small as in the network of γ' and ε phase.

A general solution of the multiphase diffusion model using numerical techniques was proposed by A. Engstrom et al. [22] and J. Andersson et al. [53]. The equation of nitrogen diffusion for gas nitriding was presented by P. Cavaliere et al. [172]. Analytical methods of nitrogen diffusion calculations were used by M. Krukovich based on Y. Lakhtin's works [214].

Gas nitriding in conditions of atmospheric pressure

Comprehensive mathematical models of the gas nitriding process at atmospheric pressure were presented by J. Ratajski [215, 216]. They allowed the structure of the nitrided alloy ferrite to be modelled as a function of time, determining the profiles of nitrogen concentration: (1) in solution, (2) bound in the nitride of the alloying elements and the amount of nitride separated and the incubation time of the γ' phase formation. They could have been used for any chemical composition of steel provided the thermodynamic data of the alloying elements were known. The nitriding model, which was based on Fick's diffusion phenomenological laws, consisted of the following expressions [216]:

$$N(x, t = 0) = 0 \tag{6.46}$$

$$-D_N^{\alpha} = \left.\frac{\partial[N(x, t < t_1)]}{\partial x}\right|_{x=0} = \beta\left\{[N]_{eq} - [N(t)]_S^{\alpha}\right\} \tag{6.47}$$

$$N(x = 0, t = t_1) = N_{\alpha/\gamma'} \tag{6.48}$$

$$N(x = 0, t \geq t_1) = N_{\alpha/\gamma'} \tag{6.49}$$

$$\left.\frac{\partial[N(x, t)]}{\partial x}\right|_{x=L} = 0 \tag{6.50}$$

$$\frac{\partial N(x, t)}{\partial t} = D\frac{\partial^2 N(x, t)}{\partial x^2} \tag{6.51}$$

$$yN(x, t) + xM(x, t) \Leftrightarrow M_x N_y \tag{6.52}$$

$$K_e = \frac{1}{[a_N]^y \cdot [a_M]^x} \tag{6.53}$$

$$K_{M_x N_y} = 1/K_e = [N(x, t)]^y \cdot [M(x, t)]^x \tag{6.54}$$

$$[N(x, t)]^y \cdot [M(x, t)]^x > K_{M_x N_y} \tag{6.55}$$

$$N(x, t) \cdot M(x, t) = K_{MN} \tag{6.56}$$

$$(M^t - \beta \cdot N^P)(N^t - N^P) = K_{MN} \tag{6.57}$$

$$N^t = N^t - N^P = N^t - \frac{N^t + \frac{M^t}{\beta} - \sqrt{\left(N^t - \frac{M^t}{\beta}\right)^2 - 4\frac{N^t M^t - K_{MN}}{\beta}}}{2} \tag{6.58}$$

$$u = M(x, t)_i \cdot N(x, t)/K_{MN}^i \tag{6.59}$$

where t_1 is the incubation time for the formation of iron nitrides γ' on the surface and L is half the thickness of the nitrided material. The detailed method of handling the model is described in the same work.

J. Ratajski proved that the problem can be solved not only with a conventional mathematical apparatus (analytical equations), but also with numerical methods (FEM) [217]. A numerical approach to modelling was also presented extensively by T. Liapina in her paper "*Phase transformations in interstitial Fe-N alloys*" [218]. At the same time, a number of researchers point to the usefulness of statistical models, in particular artificial neural networks [217, 219–221], but also evolutionary algorithms and fuzzy logic [222, 223]. The last solution is particularly appropriate in optimisation problems, in this case, in the search for optimal process parameters for given output criteria.

The creation and development of models of the conventional nitriding process was the basis for building the manual and computerised process control algorithms, a trend that had already been observed in Canada and the United States since the 1980s. Computer control of the nest for controlled gas nitriding has become possible [217].

Gas nitriding controlled with a flux of diffusing nitrogen

L. Małdziński and J. Tacikowski published a paper in 1998 concerning a new possibility to control the gas nitriding process by using simulations of the kinetics of the growth of nitrided layers [140]. In the paper, they presented the results of a computer simulation of gas nitriding, illustrating the effect of the nitrogen potential of the atmosphere NH_3/H_2 on the structure and growth kinetics of nitride phases on iron. This was a new approach to process control (optimal, from the point of view of process kinetics), the aim being to produce layers with predetermined phase structures, thicknesses of the existing zones, surface concentrations and profiles of nitrogen concentrations in the individual zones. The researchers pointed out the importance of the approach when applied to programmes of automatic control for a gas nitriding process of iron and low-carbon steels and, once the content of carbon and alloying elements is included in the calculations—also to other steels being nitrided today.

Utilitarian models for the nitriding of iron and low- and medium-alloy steels according to the *ZeroFlow* method were established between 2016 and 2021 [209]. They were used to predict the increase in the thickness of the nitrided layer and individual zones (ε, γ', $\gamma' + \varepsilon$, α), the effective thickness of the nitrided layer and distribution of the hardness profile as a function of the process parameters: duration, temperature and nitriding potential of the atmosphere (Fig. 6.8). It was possible to use them to determine process parameters based on the expected properties of the layer. However, the most important advantage of the developed models is the possibility to optimise the process parameters and thus produce the required nitrided layer in the shortest possible time and with the lowest energy and gas consumption [224]. The mathematical models of L. Małdzińki et al., established on the basis of empirical research, were used to develop the simulator. These models are based on the classical laws of thermodynamics and nitriding kinetics [209–211, 225].

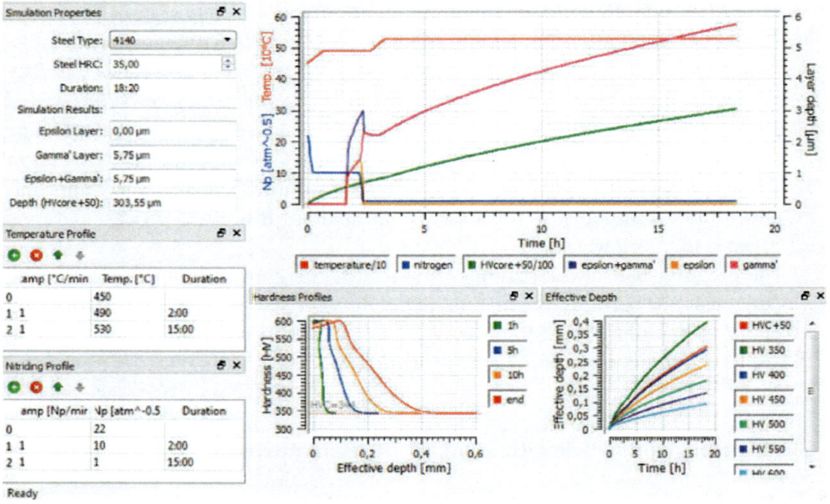

Fig. 6.8 Simulator of the kinetics of nitrided layer growth. Adapted with permission from Ref. [224] Open access under a CC BY 3.0 license, https://creativecommons.org/licenses/by/3.0/

Low-pressure nitriding

In the case of gas nitriding, the most relevant parameters influencing the growth kinetics of the nitrided layer and its phase composition are nitrogen potential, temperature, composition and flow rate of the nitriding atmosphere, the chemical composition of the nitrided material and the condition of surface preparation [216, 225–227]. For nitriding at reduced pressure, process segmentation is an important parameter determining the flux of nitrogen to the nitrided surface. Such organisation implies a division into stages of alternate saturation of the material and holding at the nitriding temperature without the presence of nitriding gas.

However, comprehensive information on the kinetics of the low-pressure nitriding process is not available in the literature. This is a process progressing in a non-stationary state, and hence solving Fick's diffusion equations describing the transport of nitrogen atoms necessary to form a favourable surface layer structure with analytical or numerical methods is challenging due to the non-linear relationship between the diffusion coefficient and the diffusion rate [228]. The problem is further complicated by the wide variety of steel and cast iron grades used (variation in chemical composition), which strongly affect the diffusion rate and phase transformations [99, 229]. The system of equations describing the relationships can only be solved if a simplified approach is taken to the phenomena surrounding the nitriding process, resulting in unacceptable discrepancies between calculations and observations. The problem can be solved using an optimisation procedure that takes into account several variable parameters, assuming that the others are fixed. A proven solution in this case is computer-aided design based on neural networks. Assumptions about the neural network model and its applications to low-pressure nitriding of tools have been presented in several papers by A. Zhechev et al. and E. Wołowiec et al. [79, 99, 196].

6.4 Thermomechanical Models

Heat transfer in the quenching process

Heating and cooling procedures are inextricably linked to any thermo-chemical process. The heating procedure is linked to the issue of uniform heat transfer in the heating chamber and the even heating of the workpieces regardless of the complexity of the shapes. The cooling process, in particular the rapid cooling called quenching, is connected with the problem of uniform heat removal and achieving a cooling rate ensuring the desired phase transformation. Quenching in particular is considered in connection with carburising (so-called *case hardening*). Critical parameters determining the properties of the product after quenching are the cooling capacity (dependent on the type of cooling medium), the chemical composition and thermal properties of the cooled component, the transformations during cooling and the resulting internal stresses. W. Dowling observed, and other researchers have confirmed, that it is important for the problem under consideration to take into account the following points [230]:

1. the effect of the chemical composition and physical properties of the material being cooled on its post-cooling properties,
2. the effect of the geometry and mass of the cooled components on the cooling rate,
3. the relationship between coefficients of thermal conductivity, material density, heat capacity and temperature,
4. the cooling rate of the component for a point in the material under consideration and its distance from the surface.

The above relationships are linked to the transient heat transfer equation (Fourier formula):

$$\rho \cdot c \cdot \frac{\partial T}{\partial t} = \frac{\partial}{\partial X}\left(K \cdot \frac{\partial T}{\partial X}\right) - \sigma_{ij} \cdot \varepsilon_{ij}^{p} + L_I \cdot \xi_I \tag{6.60}$$

where ρ—density, c—heat capacity, K—heat conduction coefficient, L—latent heat due to phase transformations, σ—stress, ε^p—degree of plastic strain, ϕ_I—volume fraction in I-th phase transformed. Second and third words of the right-hand side of Eq. (6.60) represent, respectively, the heat due to plastic deformation and the latent heat due to phase transformation. The rate of heat transfer by convection and radiation between the surface of the solid component and the flowing gas is calculated using Newton's cooling equation [231]:

$$J_Q = h \cdot S \cdot \Delta T \quad (6.61)$$

where J_Q—heat flux, S—surface area of the solid, ΔT—temperature difference between the surface of the solid and the flowing gas, h—heat transfer coefficient, which is a combination of the effects of convection and radiation.

There are a number of literature reports concerning the simulation of the hardening process using the finite element method (FEM), including the work of T. Inoue and K. Arimoto [36], D. Watt et al. [232] and R. Wallis et al. [233, 234], J. Wang et al. [235], R. Kopun et al. [236], I. Elkatatny et al. [57]. This method is available *ad hoc* in the ANSYS-CFX tool and has the advantage that a system of three equations can be solved during numerical calculations: conservation of mass, conservation of momentum and conservation of total energy [237]. A comprehensive review of the applications of the FEM in the analysis and simulation of the quenching process in more than 290 cases was given by J. Mackerle [238]. The flow of quenching gas in the quenching chamber is as much a frequently studied problem as the heat transfer in the workpiece. Practically all researchers note the usefulness of ANSYS-CFX [237] based on the finite volume method (FVM) and ANSYS-FLUENT [65] based on the finite element method (FEM) [50, 51, 54, 239–245].

The importance of the heat transfer coefficient h is similar to the diffusion coefficient D_C in carburising. It has a significant effect on the results of quenching, including thermal deformations, residual stresses and hardness distribution, hence there are a number of reports on determining its optimum value [23, 58, 62, 231, 246, 247]. According to its physical meaning, the total heat transfer coefficient depends on the surface temperature of the solid (the solid component of the system). The course of the heat transfer coefficient value can be calculated from the cooling history [62], with the inverse techniques [248] or the finite difference method [23].

Over the past two decades, various mathematical models of quenching have been developed with varying degrees of accuracy and complexity. According to J. Davis et al. [249], there are three general types of models used in the prediction of properties and computer simulation: static models, dynamic models and programmes with static and dynamic models. Static models are based on simple empirical formulae derived from experimental data and regression analysis. Dynamic models are based on solving a series of differential equations. An important category of dynamic models is the use of finite element analysis to predict deformations, residual stresses and hardness distributions [36, 232–234, 250]. Simulation of finite elements of the quenching process includes heat transfer, phase transformations and deformations. These three parts are influenced by each other and should be interconnected during the simulation process [62], which, when using finite element analysis of the gas hardening process, can be implemented using the DEFORM-HT tool [24].

Phase transformation kinetics

Phase transformations during the cooling process can lead to the regrouping of atoms within the entire volume of the quenched material, where the cooling rate is greater than or equal to the critical rate. Depending on whether diffusion occurs during the transformation, phase transformations during quenching can be classified as diffusion and non-diffusion transformations, i.e. martensitic, transformations. The transformation of the diffusion phase includes the diffusion of carbon/nitrogen and other atoms of the dissolved substance and depends on time [62].

The transformation of the diffusion phase comprises transformations from austenite to ferrite and cementite. The kinetics of the isothermal transformations of ferrite/perlite and bainite are modelled according to the law developed by Johnson-Mehl-(JMA) [25]. It is used to predict the volume fraction of transformed phases [26, 251]:

$$\phi_k = 1 - \exp(-b \cdot t^n) \tag{6.62}$$

where ϕ_k—the volume fraction of the transformed austenite, b and n are determined from the experimental data of the isothermal phase transformation, t—duration of the phase transformation. k = 2 for ferrite/perlite and k = 3 for bainite. The volume fraction of the component in continuous cooling can be obtained by means of a mathematical transformation. The volume fraction of the martensite phase is calculated according to the formula [23]:

$$\phi_M = (1 - \phi_1 - \phi_2) \cdot \left[1 - \left(\frac{T - M_f}{M_s - T}\right)^{2.5}\right] \tag{6.63}$$

where M_s and M_f are the start and end of the martensitic transformation, ϕ_M is the volume fraction of the martensite phase. The above value can also be calculated from the equation of C. Magee [252]:

$$\phi_M = 1 - \exp(\psi_1 \cdot T + \psi_2) \tag{6.64}$$

where ϕ_M—volume fraction of the martensite phase, T—temperature, ψ_1 and ψ_2—constants determined from the initial and final temperatures of martensite transformations.

During the quenching process of a steel component, both diffusion and martensite phase transformations can occur simultaneously at different locations in the component, hence the accuracy of the phase transformation models is very important for the quenching simulation results. Phase transformations can be conveniently modelled with tools using finite element models, as mentioned above. This principle was confirmed by A. Sugianto et al. [59] who used the available DEFORM 2D and 3D tool to study the predicted kinetics of phase transitions during heating, carburising, diffusion and quenching processes.

H. Cheng et al. report that heat conduction during quenching is a non-linear problem, and the temperature calculation is closely related to the surface heat transfer coefficient and phase transformation. The authors highlight the advantages of the finite element method for the mathematical modelling of martensite in continuous cooling and for calculating the temperature field of a component during high-pressure gas quenching [23].

The properties of steel are intrinsically related to phase transformations during quenching. Accurate prediction of the mechanical properties of quenched steel components has been considered by many researchers [50, 58, 238]. A new trend has been observed now in employing data mining methods (especially neural network models) to predict the properties of a quenched component. M. Kianezhad et al. [103] investigated the use of neural networks (NN) to predict the hardness of steel parts quenched in different hardening media. They demonstrated that neural networks are an excellent method for accurately predicting the hardness of steel parts and can be used to predict accurate hardness for the entire range of harnesses and different phases formed in quenched steel parts.

B. Podgórnik et al. [253] found that neural modelling can be successfully used for detailed and parallel analyses of various relevant parameters, as well as for optimising the vacuum heat treatment process of hot work tool steels depending on their composition. Their neural network of the type MLP (Multi-Layer Perceptron) with a back-propagation error learning scheme, as well as a scheme of four layers of neurons, was based on experimentally obtained tempering diagrams of ten different hot work tool steel compositions.[8]

Stress and deformation analysis

According to H. Cheng et al., in order to obtain the best thermal stresses, deformations during gas quenching and residual stresses after gas quenching when determining the surface heat transfer coefficients, the coupled effect between temperatures and phase transformations should be taken into consideration [23, 255]. Consequently, the deformation resulting from the quenching process mainly stems from the thermal gradient and phase transformations in the component. In the case of deformation, incremental deformation is assumed to consist of several members:

$$\Delta\varepsilon = \Delta\varepsilon^{e} + \Delta\varepsilon^{p} + \Delta\varepsilon^{\mathrm{th}} + \Delta\varepsilon^{\mathrm{tr}} + \Delta\varepsilon^{\mathrm{tp}} \tag{6.65}$$

where upper indices e, p, th, tr and tp represent, respectively, elastic transformation ε^{e}, plastic transformation ε^{p} (plastic strain), thermal transformation $\varepsilon^{\mathrm{th}}$, phase transformation strain $\varepsilon^{\mathrm{tr}}$ and transformation plasticity strain $\varepsilon^{\mathrm{tp}}$. Phase transformations and transformation plasticity make the quenching process a highly non-linear problem [62]. Calculation formulae for ε^{e}, ε^{p} and $\varepsilon^{\mathrm{th}}$ were described by P. Ferro, M. Dalgic et al. [256, 257].

[8] Until a certain point in the state of the art in computer science, a network architecture above two hidden layers was considered too complex. According to G. Cybenko, one hidden layer should be sufficient to transform any continuous function [254]. However, solutions based on networks with a multilayer structure are now increasingly being adopted (*Deep Learning* trend), even though their design is more difficult than with simple networks.

The plasticity of the transformation is also often calculated as:

$$\varepsilon^{\text{tp}} = K \cdot \sigma \cdot f(w) \tag{6.66}$$

where σ—the applied stress, K—the plasticity parameter of the transformation, and $f(w)$—a formula describing the evaluation of the plasticity of the transformation as a function of the amount in the formed phase of [257]. According to P. Ferro et al. [256], the plasticity of phase transformations can be defined as anomalous plastic deformation that can be observed. Where the phases are perfectly plastic, the exact expression describing the deformation of plastic transformation is given in the paper by J. Leblond et al. [258]:

$$\varepsilon^{\text{tp}} = \begin{cases} 0, & \text{for } p \leqslant 0.03 \\ -\frac{3\Delta\varepsilon^{\text{th}}_{i\to j}}{\sigma^{y}_{i}} \text{Sh}\left(\frac{\sigma^{\text{eq}}}{\sigma^{y}}\right)(\ln(p))\dot{p}, & \text{for } p > 0.03 \end{cases} \tag{6.67}$$

where p—volume fraction of j-th phase (created by transformation), σ^{eq}—macroscopic von Mises stress, σ^{y}—macroscopic yield stress evaluated either by a simple linear mixture rule or by a more accurate non-linear mixture rule [258, 259], S—the macroscopic deviator of stress, $\Delta\varepsilon^{\text{th}}_{i\to j}$—the difference in thermal deformations between two phases ($\varepsilon^{\text{th}}_{j} - \varepsilon^{\text{th}}_{i}$), σ^{y}_{i}—the yield stress of i-th phase, the function $h(\sigma^{\text{eq}}/\sigma^{y})$ is defined as follows [256]:

$$h\left(\frac{\sigma^{\text{eq}}}{\sigma^{y}}\right) = \begin{cases} 1, & \text{for } \frac{\sigma^{\text{eq}}}{\sigma^{y}} \leqslant \frac{1}{2} \\ 1 + 3.5\left(\frac{\sigma^{\text{eq}}}{\sigma^{y}} - \frac{1}{2}\right), & \text{for } \frac{\sigma^{\text{eq}}}{\sigma^{y}} > \frac{1}{2} \end{cases} \tag{6.68}$$

The assumption of $p = 0.03$ (3%) takes into account the fact that the analytical formulation of transformation plasticity (due to the assumption that the entire austenitic phase undergoes plasticity from the onset of ideal plasticity) overestimates the magnitude of the phenomenon for very low values of p, where the strain is essentially elastic [260]; while the function (6.68) considers the non-linearity with respect to the applied stress, which only becomes important when the stress is of the order of σ^{y} [260]. E. Mittemeijer and M. Somers referred to the deformation of phase transformations as [261]:

$$\varepsilon^{\text{tr}} = \sum_{i} y_{\phi_i} \varepsilon^{\text{tr}}_{\phi_i} \tag{6.69}$$

where y_{ϕ_i} is the change over time in the volume fraction of the ϕ_i-th phase, and $\varepsilon^{\text{tr}}_{\phi_i}$ is the deformation rate of the phase transformation due to precipitation/dissolution of the ϕ_i-th phase.

D. Ju et al. [47] inform that the total strain factor ε after heat treatment, including carburising or nitriding, is the sum of the elastic strain factor ε^{e}, plastic strain factor ε^{p}, thermal strain factor ε^{th}, transformation strain factor ε^{tr} and diffusion strain factor

ε^{df} of carbon and nitrogen. The authors also provide mathematical relationships that can be used to calculate these values for the cooled component.

Most researchers report that they use the finite element method (FEM) or response surface method (RSM) in their studies to determine deformations (strains, quenching deformations), residual stresses, average surface hardness and the standard deviation of surface hardness [57, 61, 62, 245, 246, 262, 263]. The literature of the topic also describes applications of the finite element method for predicting residual stresses and minimising distortions during heat treatment of steels [46, 61, 245, 264]. Some commercial simulators, such as HEARTS [265], SYSWELD [38], DEFORM-HT [255], DANTE [266, 267] i COSMAP [59], have been adapted and are used by the steel parts manufacturing industry to produce automobiles, machinery and tools [268].

Non-metallic precipitations in the grains (carbides, carbonitrides), which can harden the material, are also important for the material properties. A thermodynamic model for precipitation of carbonitrides in high strength low-alloy steels (HSLA) was presented by H. Adrian using numerical modelling, including generalised Newton method and cellular automata method [269, 270, 270–272].

6.5 Optimisation Models

The greatest advantage of using computational techniques in the aspect of the development of vacuum thermo-chemical treatment is probably that they can be used in optimisation procedures. The number of optimisation problems solved numerically is steadily growing and the choice of the computational method strictly depends on the individual characteristics of the problem. Three issues are given below using low-pressure carburising as an example.

Optimisation of carbon flux in the carburising process

The process of low-pressure multi-segment carburising (LPC) takes place under conditions of the excess carburising atmosphere in the carburising chamber [160, 161, 273]. Historically, this results from the difficulty in estimating the amount of carbon required, caused by the variability of steel absorption during the process. In addition, the gas introduced into the carburising chamber dissociates only partially and the degree of this dissociation depends on the temperature and time of staying in the chamber. A method of calculating the flow rate of the carburising gas for acetylene was proposed by H. Antes, noting that the procedure required knowledge of a curve of a fraction of acetylene dissociated in temperature and time function [274], but once calculated, the flow rate remained constant for all saturation cycles. Meanwhile, the absorptivity of the carburised material decreases as the carbon concentration gradient in the material decreases, so it is important to match the flow rate individually to the instantaneous demand for carbon and to ensure maximum utilisation of the carbon (reduction of the excess) supplied by the carbon-bearing gas at each stage of the process. At the same time, the algorithm must ensure that sufficient carbon is supplied to the surface so that the carbon flux entering the surface does not become a parameter limiting the rate of the process.

The relationship described by Fick's first law indicates that the demand for carbon for low-pressure carburising is lowering in proportion to the growth in the carbon content of the material being carburised. A mathematical analysis of Fick's law leads to the conclusion that the saturation (boost) stages should use individual flows of carburising gases according to the demand for the mixture in a given stage. The size of this demand is described by the relationship [275]:

$$F_{\text{boost}} = \frac{22.4}{m_C} \cdot \frac{7.87 \cdot 10^6 \cdot \int(\partial C \partial x)}{t_b} \cdot S \tag{6.70}$$

where F_{boost}—flow of the carburising mixture in the saturation segment, m_C—amount of carbon in 1 mole of the carburising mixture, t_b—duration of saturation segment, S—area of the carburised surface, $\int(\partial C \partial x)$—amount of carbon absorbed by the surface during the *boost-diffusion* cycle. Since the mixture in the carburising chamber dissociates only partially, the actual demand for the carbon-bearing mixture in the saturation stage is the ratio of the calculated value (Eq. 6.70) and the degree of dissociation of the mixture, determined according to the H. Antes procedure [274]. The value of flow reduction depends on the temperature and the way the process is organised. In a comparative study of two LPC processes presented in the paper by E. Wołowiec-Korecka et al. (Fig. 6.9), the amount of mixture used was reduced by 46% as described, keeping the values of all parameters of the carburised layer identical [275].

The integral $\int(\partial C \partial x)$ denoting the amount of carbon absorbed by the surface of the iron alloy during the saturation-heating cycle in a reasonable time dimension can only be solved by numerical calculations, hence efficient optimisation of the low-pressure carburising process in terms of flow became possible only after

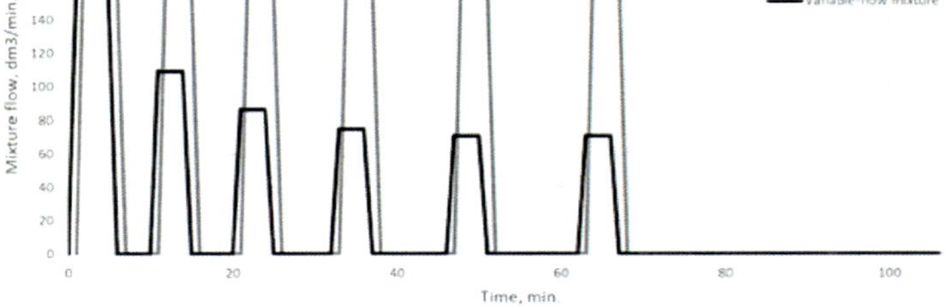

Fig. 6.9 Comparison of consumption rates of carbon-bearing gases in LPC processes with constant and variable flow of carbon-bearing mixture: LPC process at 960°C, CD_{Eff}: 0.8 mm. Reprinted with permission from Ref. [275] Open access under a CC BY 4.0 license, https://creativecommons.org/licenses/by/4.0/

computational tools became widespread. In terms of transferring the results of the calculations to the control module of the device, this has been possible since PLCs have become popular.

Intensification of the carburising process by surface phenomena—control of surface phenomena

The alloying elements Cr, W, Mo, V, Ti, Nb and Fe present in the steel may cause the precipitation of carbides with a simple and/or complex structure during both carburising and nitriding [276]. When heating components for heat treatment, carbides dissolve into austenite, altering the kinetics of this phase transformation, as well as increasing the hardenability during cooling. The cementite present in the structure of the material before treatment also decomposes and becomes a source of carbon atoms [277]. In the subsequent stages of treatment, during the cooling phase, carbide-forming elements from dissolved carbides become part of the martensite (or perlite or bainite) structure or form new carbides. The functional consequences of the presence of carbides in iron alloys can be both positive and negative. The fine-grained, spherical carbides that are released in the volume of grains strengthen the material. However, the appearance of a cementite lattice brings with it a number of adverse effects: local decarburisations appear in the material [278]; the percentage of alloying elements increases; where carbides are present, the material becomes harder and more brittle [279], tends to crack and becomes more susceptible to corrosion [278]. Alloying elements are important for the properties of steels, also for the phenomena of (carbide- and nitride-forming) precipitation. The thermodynamic aspects of precipitations of carbides and nitrides have been extensively described in the literature, both from theory and by numerical simulations [269, 280].

At the same time, considering the process from a theoretical point of view, the fastest way to carry out the process is to run the process with the size of the carbon flux ensuring that a concentration equal to the maximum carbon concentration in the austenite at the process temperature is achieved on the surface of the steel. In particular, multi-segment low-pressure carburising is privileged with such organisation. When the surface concentration of carbon decreases during the annealing stage, the subsurface carbide precipitates continue to be a source of carbon supplying the diffusion process. It is therefore optimal to carry out the carburising process under technological parameters favourable to carbide-forming processes, under the condition that all precipitates are diffused before the end of the process. This issue can be solved with analytical-numerical methods [160], but the optimum organisation of the low-pressure carburising process (organisation of the process into saturation-heating stages) requires more complex optimisation methods like neural networks [100, 281].

Multi-stage and correction processes

Not each thermo-chemical treatment process is progressing in an exemplary way. Apart from human error, equipment failure or disruption of technological conditions can occur, resulting in product properties different from those expected. In many

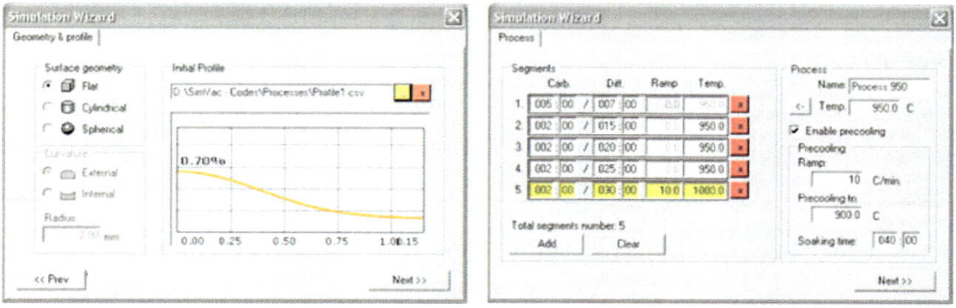

Fig. 6.10 Low-pressure carburising simulator: **a** window for the definition of the surface geometry of the surface layer and its initial carbon concentration profile, **b** window for setting the times and temperatures of the repair process segmentation

cases, this means rejecting the product from further production. In some cases, however, it is possible to establish a "remedy process". The simulation algorithm is able to reproduce the process flow, determine the product properties after the process (Fig. 6.10a) and propose a procedure (second process), bringing the component properties to the desired values. Figure 6.10 shows the situation of the component after faulty carburising. The properties were reproduced and the "missing" part of the process was determined in the simulation. E. Wołowiec et al. showed that it is possible to develop a continuation process even without knowledge of the initial process, based on an experimentally measured carbon concentration profile in the product [282].

6.6 Monitoring of Phenomena in Real Time

Apart from predictive issues of the phenomena, another important issue is the monitoring and management of their progress. In the utilitarian aspect, this means being able to track the thermal-diffusion process as it occurs and logistically manage its start and end.

A model of the phenomenon needs to be built to monitor the course of the phenomenon, as described in the previous subsections, and additional modelling methods are not required. A separate issue is the communication with the device's sensors and the computational complexity of the algorithms. Communication with the device is implemented through the connection to the device's PLC and is an issue in the field of automation, developed since the beginning of the third industrial revolution. The computational complexity of modelling algorithms is a computer science issue. Many models constructed with numerical methods (FEM, FVM, FDM) using off-the-shelf computational tools are computationally time-consuming, which translates into days and weeks of computations, so real-time computing has for many years been a trade-off between computational accuracy and computational complexity. Nowadays, developments in computer processing power allow improvements in both aspects with each new generation of computers. From a technical point of view, the monitoring of the process taking place in the heat treatment chamber is as follows:

1. connection to the PLC and retrieval of the values of actual process parameter recorded by the device's sensors,
2. retrieval of information from the PLC about the current duration of the actual process,
3. calculation of the surface layer properties of the workpiece material at the current time instant,
4. display of calculation results for verification by the supervising user (reporting),
5. if the user does not react—return to step 1,
6. if the user does react—retrieving the expected layer properties and determining new values of process parameter for the material with properties at the current time instant,
7. communication with the PLC and transmission of new values of parameters.

Issues related to real-time monitoring of carburising and gas quenching phenomena were described by E. Wołowiec et al., while issues related to gas nitriding were presented by J. Michalski et al. (Fig. 6.11) and L. Małdziński et al. [141, 224, 282, 283].

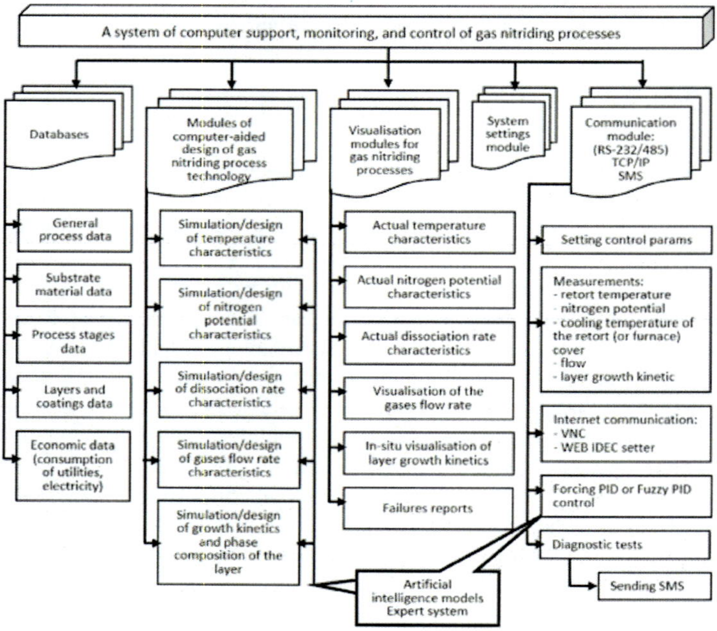

Fig. 6.11 Module structure of system of gas nitriding. Reprinted with permission from Ref. [141] Copyright (2008) Sigma-NOT

6.7 Predictive Maintenance

In industrial practice (commercial quenching, manufacturing companies), treatment processes are running in parallel and the economic use of all resources needs to be planned. The concept of maximising the use of heat treatment equipment and organising it into efficient production lines (lean manufacturing) has appeared in the literature for years. With the concept of the fourth industrial revolution, the issue of predictive maintenance (PM) has gained importance [284]. In traditional preventive maintenance service, service inspections are conducted according to a predetermined schedule. As a result, the unit is out of service at the time of inspection and replacement of consumable components regardless of their condition. On the other hand, extending maintenance intervals carries the danger of a major breakdown, meaning even more costs and wasted time. Predictive maintenance assumes to perform maintenance works when there is a clear indication of the impending failure. This moment can be captured properly based on one of three ways (Fig. 6.12): (1) estimation of the condition of the device, (2) analysis of the physical model of the process and its impact on the device, (3) a data-driven approach (statistical model). The rules and patterns obtained with statistical methods and machine learning methods are used in the third approach (ML) [284–286].

The classic analytical approach of *Predictive Maintenence* considers: event sequence analysis, deviation from the pattern and failure-free operation time models. An analysis of the sequence of events looks for sequences of events that precede failures (e.g. motor overload, high machine vibrations). In the analysis of deviations from the pattern, data instances that deviate from a properly functioning device are analysed and the distance of the current state from the pattern prior to a specific

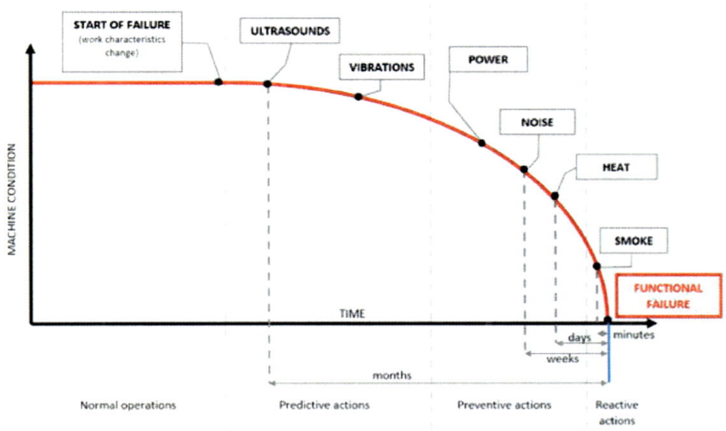

Fig. 6.12 Diagram of device status on a time scale, with time periods for predictive, preventive and reactive actions marked

failure is measured, i.e. data on the past occurrence of a specific failure and the behaviour of the device prior to its occurrence. A multivariate data analysis and the criterion of distance from the pattern are applied when examining deviations from a pattern. This allows previously unobserved failures to be predicted, which is a great advantage of this approach; it also allows wear and tear to be monitored and signalled. The analytical methods applied comprise teacher-free learning methods (unsupervised learning), visualisation, control cards, multivariate analyses (including analysis of principal components, segmentation, cluster analysis), neural networks (including the traditional Kohonen network or self-organising network) and deep neural networks.

When analysing patterns prior to a specific change, training patterns are used (typically several hundred courses are needed for a training set), including learning with a teacher, pattern classification, regression, logistic regression, neural networks, SVM, boosted trees and random forest. Classical analysis of time series and regression (Cox regression and regression models) are used in failure-free operation time models.

Predictive Maintenence is an obvious element of the Industry 4.0 and Internet of Things (IoT) strategies. The proliferation of this approach is influenced by the low costs of sensors, the ease of retrieving, transmitting and collecting data from machines and equipment, and the proliferation of complex analytical tools.

6.8 Further Directions of Development

Computer-aided modelling of phenomena and material features is becoming a common tool, and modern progress in materials engineering is closely associated with the application and development of computation methods. Computer-aided reproduction of the processes and phenomena helps to predict the properties of materials with a considerable reduction of the financial outlays and the time needed for experiments and implementation. The near future will see an increasing demand for effective, precise and easy-to-handle computation tools for predicting and designing thermal and thermo-chemical low-pressure treatments. Both conventional and numerical models, as well as those of the data mining type, are already widely applied in modelling such processes.

Conventional models based on analytical equations undoubtedly yield the most accurate results. However, the constant growth of the computation capacity of computers favours the development of numerical methods. It is expected that results will be obtained by numerical methods more quickly and more accurately. This process is favoured by the development of knowledge of low-pressure treatments, which enables to establish of the coefficients describing the process in individual devices with greater precision.

An increase in the body of knowledge will be accompanied by an increase in the amount of data regarding LPC, LPN and HPGQ processes, and the need for effective analysis, faster searches and acquiring of new knowledge from data are going to lead to an increase in the applications of DM methods.

Due to their nature, some phenomena are still going to be too complicated, or their modelling by conventional and numerical methods will be time consuming. Moreover, new applications of thermal treatment will emerge, with unelucidated and unexplored phenomena (low-pressure thermal treatment of sinters, low-pressure thermal treatment of 3D-printed elements, etc.). It seems that methods of artificial intelligence will be used first, both with regard to predictions (artificial neural networks) and optimisation (genetic and evolutionary algorithms). However, teaching databases will increasingly help to improve the precision of their predictions [1].

References

1. E. Wołowiec-Korecka, Modeling methods for gas quenching, low-pressure carburizing and low-pressure nitriding. Eng. Struct. **177**, 489–505 (2018)
2. M. Amanowicz, W. Gasparski, *Design Studies: Elements of Design Knowledge* (WNT, Warsaw, 1988)
3. Y. Lakhtin, Y. Kogan, A. Bulgach, Prospects for computer usage in heat treatment and case hardening. Metal Sci. Heat Treat. **26**(1), 3–9 (1984)
4. M. Berry, G. Linoff, *Data Mining Techniques: For Marketing, Sales and Customer Support* (John Willey & Sons, Hoboken, 1997)
5. L. Kukiełka, *Fundamentals of Engineering Research* (PWN, Warsaw, 2002). (in Polish)
6. S. Weiss, N. Indurkhya, *Predictive Data Mining* (A Practical Guide (Morgan Kaufmann Publisher, Burlington, 1998)
7. D. Braha, *Data Mining for Design and Manufacturing* (Methods and Applications (Kluwer Academic Publisher, Heidelberg, 2001)
8. N. Draper, H. Smith, *Applied Regression Analysis* (Wiley-Interscience, New York, 1998)
9. P. Januszewski, *Practical Applications of Regression Techniques (Tutorial)* (StatSoft Poland, Cracow, 2021). (in Polish)
10. S. Malinov, W. Sha, Software products for modelling and simulation in materials science. Comput. Mater. Sci. **28**(2), 179–198 (2003)
11. P. Kula, R. Pietrasik, K. Dybowski, Vacuum carburizing-process optimization. J. Mater. Process. Technol. **164–165**, 876–881 (2005)
12. A. Ochsner, J. Gegner, G. Mishuris, Effect of diffusivity as a function of the method of computation of carbon concentration profiles in steel. Metal Sci. Heat Treat. **46**, 148–151 (2004)
13. M. Semenov, A. Smirnov, M. Ryzhova, Computation of carbon concentration curves in vacuum carburizing of steels. Metal Sci. Heat Treat. **55**(1–2), 38–42 (2013)
14. J. Goldstein, A. Moren, Diffusion modeling of the carburization process. Trans. Metall. Soc. AIME **9A**, 1515–1525 (1978)
15. H. Jimenez, M. Staia, E. Puchi, Mathematical modeling of a carburizing process of a SAE 8620H steel. Surf. Coat. Technol. **120**, 358–365 (1999)
16. M. Semenov, A. Smirnov, R. Fakhurtdinov, O. Ospennikova, V. Gromov, Optimization of modes of vacuum carburizing of gears from heat-resistant steel VKS-7 on the basis of computational design. Metal Sci. Heat Treat. **57**(1–2), 28–31 (2015)
17. A. Wells, W. Batz, R. Mehl, Diffusion coefficient of carbon in austenite. Trans. Metall. Soc. AIME **188**, 553–560 (1950)
18. L. Kaufman, S. Radcliffe, M. Cohen, Decomposition of austenite by diffusional processes, in *Proceedings of a Symposium Held in Philadelphia, Pennsylvania, October 19, 1960*, eds. by V. Zackay, H. Aaronson (Chichester, 1962)

19. G. Leyens, G. Woelk, J. Wuenning, Berechnung der Aufkohlung nach dem Saettigungs-Ausgleichs-Verfahren. Archiv für das Eisenhüttenwesen **47**, 385–390 (1976). (in Polish)
20. G. Tibbets, Diffusity of carbon in iron and steels at high temperatures. J. Appl. Phys. **51**, 4813–4816 (1980)
21. J. Morral, B. Dupen, C. Law, Application of commercial computer codes to the modelling carburizing kinetics of alloy steels. Metall. Trans. A **23**, 2069–2070 (1992)
22. A. Engström, L. Höglund, J. Ågren, Computer simulation of diffusion in multiphase systems. Metall. Mater. Trans. A **25**(6), 1127–1134 (1994)
23. H. Cheng, J. Xie, J. Li, Determination of surface heat-transfer coefficients of steel cylinder with phase transformation during gas quenching with high pressures. Comput. Mater. Sci. **29**(4), 453–458 (2004)
24. Scientific Forming Technologies, *DEFORM Users Manual* (Scientific Forming Technologies Corporation, Columbus, 1999)
25. W. Johnson, F. Mehl, Reaction kinetics in processes of nucleation and growth. Trans. Metall. Soc. AIME **135**, 416–425 (1939)
26. R. Price, A. Fletcher, Determination of surface heat-transfer coefficients during quenching of steel plates. Metals Technol. **7**(1), 203–211 (1980)
27. D. Knoke, P. Burke, *Log-linear Models* (SAGE Publications, Indiana University, 1980)
28. A. Stanisz, *Log-linear Analysis (Tutorial)* (StatSoft Poland, Cracow, 2021). (in Poland)
29. A. Eve, E.-Y. Mun, *Log-linear Modeling* (Concepts, Interpretation and Application (Wiley, Hoboken, 2013)
30. A. Agresti, *An Introduction to Categorical Data Analysis* (Wiley, Hoboken, 2019)
31. R. Christensen, *Log-linear Models and Logistic Regression* (Springer, New York, 2013)
32. B. Ratner, *Statistical Modeling and Analysis for Database Marketing* (Chapman & Hall, London, 2003)
33. A. Luszniewicz, *Applications of Statistical Methods* (National Economic Publishing House, Warsaw, 1972). (in Polish)
34. M. Dobosz, *Computer-aided Statistical Analysis of Test Results* (Academic Publishing House EXIT, Warsaw, 2001). (in Polish)
35. D. Kim, H. Cho, W. Lee, K. Cho, Y. Cho, S. Kim, H. Han, A finite element simulation for carburizing heat treatment of automotive gear ring incorporating transformation plasticity. Mater. Design **99**, 243–253 (2016)
36. T. Inoue, K. Arimoto, Development and implementation of CAE system HEARTS for heat treatment simulation based on metallo-thermomechanics. J. Mater. Eng. Perform. **6**, 51–60 (1997)
37. M. Yaakoubi, M. Kchaou, F. Dammak, Simulation of heat treatment and materials with the use of the Abaqus software. Metal Sci. Heat Treat. **55**(7–8), 386–392 (2013)
38. Southwest Research Institute, *SYSWELD. A Predictive Model for Heat Treat Distortion* (Southwest Research Institute, San Antonio, 1992)
39. D. Pont, T. Guichard, Sysweld: Welding and heat treatment modelling tools, in *Computational Mechanics'95*. ed. by S. Atluri, G. Yagawa, T. Cruse (Springer, Berlin, Heidelberg, 1995), pp. 248–253
40. B. Ferguson, G. Petrus, T. Pattok, A software tool to simulate quenching of alloy steels, in *Proceedings of the 3rd International Conference on Quenching and Control of Distortion* (Materials Park, 1999), pp. 188–200
41. D. Ju, Y. Ito, T. Inoue, Simulation and verification of residual stresses and distortion in carburizing–quenching process of a gear shaft, in *Proceedings of the 4th International Conference on Quenching and Control of Distortion* (Materials Park, 2003), pp. 291–296
42. Wilde & Partners, Virtual heat treating using DEFORM-HT. Aircr. Eng. Aerosp. Technol. **70**(4) (1998)
43. K. Arimoto, D. Lambert, G. Li, A. Arvind, W. Wu, Development of heat treatment simulation system DEFORMTM-HT, in *Proceedings of the 18th Conference on Heat Treating*, eds. by R. Wallis, H. Walton (Materials Park, 1998), pp. 639–654

44. C. Liu, X. Xu, Z. Liu, A FEM modeling of quenching and tempering and its application in industrial engineering. Finite Elem. Anal. Design **39**(11), 1053–1070 (2003)
45. Z. Li, B. Ferguson, V. Nemkov, R. Goldstein, J. Jackowski, D. Fett, Modeling distortion and residual stresses of an induction hardened truck axle. Adv. Mater. Process. **1**(1), 62–64 (2013)
46. S.-J. Lee, D. Matlock, C. van Tyne, Comparison of two finite element simulation codes used to model the carburizing of steel. Comput. Mater. Sci. **68**, 47–54 (2013)
47. D. Ju, C. Liu, T. Inoue, Numerical modeling and simulation of carburized and nitrided quenching process. J. Mater. Process. Technol. **143–144**, 880–885 (2003)
48. D. Buchholz, R. Khan, S. Bajohr, R. Reimert, Computational fluid dynamics modeling of acetylene pyrolysis for vacuum carburizing of steel. Ind. Eng. Chem. Res. **49**(3), 1130–1137 (2010)
49. R. Khan, S. Bajohr, D. Buchholz, R. Reimert, H. Minh, K. Norinaga, V. Janardhanan, S. Tischer, O. Deutschmann, Pyrolysis of propane under vacuum carburizing conditions: An experimental and modeling study. J. Anal. Appl. Pyrol. **81**(2), 148–156 (2008)
50. N. Lior, The cooling process in gas quenching. J. Mater. Process. Technol. **155–156**, 1881–1888 (2004)
51. T. Wang, X. Yu, T. Chen, Q. Zhang, *CFD simulations of quenching process for partial oxidation of methane: comparison of jet-in-cross-flow and impinging flow configurations* (Chin. J. Chem, Eng, 2016)
52. K. Yada, O. Watanabe, Reactive flow simulation of vacuum carburizing by acetylene gas. Comput. Fluids **79**, 65–76 (2013)
53. J. Andersson, T. Helander, L. Höglund, P. Shi, B. Sundman, Thermo-Calc & DICTRA, computational tools for materials science. Calphad **26**(2), 273–312 (2002)
54. K. Babu, T.S. Kumar, Mathematical modelling of surface heat flux during quenching. Metall. Mater. Trans. B **41**, 214–224 (2010)
55. P. Kula, R. Atraszkiewicz, E. Wołowiec, Modern gas quenching chambers supported by SimVaC Plus Hardness application. Ind. Heat. **3**, 55–58 (2008)
56. R. Collin, S. Gunnarson, D. Thulin, A mathematical model for predicting carbon concentration profiles of gas-carburized steel. J. Iron Steel Inst. **210**(10), 785–789 (1972)
57. I. Elkatatny, Y. Morsi, A. Blicblau, S. Das, E. Doyle, Numerical analysis and experimental validation of high pressure gas quenching. Int. J. Therm. Sci. **42**(4), 417–423 (2003)
58. J.E. Pritchard, G. Nurnberg, M. Shoukri, Computer modeling of pressure gas quenching in vacuum furnaces. Heat Treat. Metals **4**, 79–83 (1996)
59. A. Sugianto, M. Narazaki, M. Kogawara, A. Shirayori, S. Kim, S. Kubota, Numerical simulation and experimental verification of carburizing-quenching process of SCr420H steel helical gear. J. Mater. Process. Technol. **209**(7), 3597–3609 (2009)
60. Z. Li, A.M. Freborg, B.D. Hansen, T.S. Srivatsan, Modeling the effect of carburization and quenching on the development of residual stresses and bending fatigue resistance of steel gears. J. Mater. Eng. Perform. **22**(3), 664–672 (2013)
61. O. Karabelchtchikova, I. Rivero, S. Hsiang, Modeling of residual stress distribution in D2 steel via grinding dynamics using a second-order damping system. J. Mater. Process. Technol. **198**, 313–322 (2008)
62. Z. Li, R.V. Grandhi, R. Srinivasan, Distortion minimization during gas quenching process. J. Mater. Process. Technol. **172**(2), 249–257 (2006)
63. K. Bongartz, W. Quadakkers, R. Schulten et al., A mathematical model describing carburization in multielement alloy system. Metall. Trans. A **20**, 1021–1028 (1989)
64. J. Crank, *The Mathematics of Diffusion* (Oxford University Press, New York, 1979)
65. Fluent Inc., *Fluent 6.3 User's Guide* (Fluent Inc., Lebanon, 2006)
66. G. Piatetsky-Shapiro, W. Frawley, *Knowledge Discovery in Databases* (MIT Press, 1991)
67. U. Fayyad, G. Piatetsky-Shapiro, P. Smyth, R. Uthurusamy, *Advances in Knowledge Discovery and Data Mining* (MIT Press, Cambridge, 1996)
68. J. Han, M. Kamber, *Data Mining: Concepts and Techniques* (Academic Press, Waltham, 2001)
69. G. Migut, T. Demski, *Data Minig I (Tutorial)* (StatSoft Poland, Cracow, 2012). (in Polish)

70. S. Serengil, Evolution of neural networks and deep learning|sefiks.com. https://sefiks.com/2017/10/14/evolution-of-neural-networks/ (2017). Accessed 28 Jan 2024
71. I. Aziz, Fundamental knowledge of machine learning. https://medium.com/analytics-vidhya/fundamental-omachine-learning-ada28afa1bd3 (2019). Accessed 24 Jan 2024
72. R. Tadeusiewicz, *Artificial Intelligence Archipelago* (EXIT, Warsaw, 2021). (in Polish)
73. W. Duch, J. Korbicz, *Neural Networks* (Academic Publishing House EXIT, Warsaw, 2000). (in Polish)
74. D. Gianola, H. Okut, K.A. Weigel, G.J. Rosa, Predicting complex quantitative traits with bayesian neural networks: A case study with Jersey cows and wheat. BMC Genet. **12**, 87 (2011)
75. S. Malinov, W. Sha, Z. Guo, Application of artificial neural network for prediction of time-temperature-transformation diagrams in titanium alloys. Mater. Sci. Eng. A **283**(1), 1–10 (2000)
76. K. Genel, Use of artificial neural network for prediction of ion nitrided case depth in Fe-Cr alloys. Mater. Design **24**(3), 203–207 (2003)
77. S. Malinov, W. Sha, Application of artificial neural networks for modelling correlations in titanium alloys. Mater. Sci. Eng. A **365**(1–2), 202–211 (2004)
78. L.A. Dobrzański, J. Trzaska, Application of neural networks for the prediction of continuous cooling transformation diagrams. Comput. Mater. Sci. **30**(3–4), 251–259 (2004)
79. A. Zhecheva, S. Malinov, W. Sha, Simulation of microhardness profiles of titanium alloys after surface nitriding using artificial neural network. Surf. Coat. Technol. **200**(7), 2332–2342 (2005)
80. W. Sha, K.L. Edwards, The use of artificial neural networks in materials science based research. Mater. Design **28**(6), 1747–1752 (2007)
81. X. Liujie, X. Jiandong, W. Shizhong, Z. Yongzhen, L. Rui, Optimisation of chemical composition of high speed steel with high vanadium content for abrasive wear using an artificial neural network. Mater. Design **28**(3), 1031–1037 (2007)
82. X. Liujie, X. Jiandong, W. Shizhong, P. Tao, Z. Yongzhen, L. Rui, Artificial neural network prediction of heat-treatment hardness and abrasive wear resistance of high-vanadium high-speed steel (HVHSS). J. Mater. Sci. **42**(8), 2565–2573 (2007)
83. E. Wołowiec, Application of artificial intelligence methods for modeling and simulation of vacuum carburizing processes of steel, in *Application of Statistical Methods in Scientific Research*, vol. IV (StatSoft Poland, Cracow, 2012), pp. 433–444. (in Polish)
84. M. Hagan, H. Demuth, M. Beale, *Neural Networks Design* (PWS Publishing Company, Boston, 1996)
85. E. Wołowiec, *The Application of Artificial Intelligence Methods in Development and Technical Realization of Surface Engineering Processes*. PhD thesis (Lodz University of Technology, Lodz, 2009). (in Polish)
86. E. Wołowiec, *Computer Design of Heat Treatment Processes* (Lodz University of Technology, Lodz, 2013). (in Polish)
87. S. Saha, Z. Gan, L. Cheng, J. Gao, O. Kafka, X. Xie, H. Li, M. Tajdari, H. Kim, W. Liu, Hierarchical deep learning neural network (HiDeNN): An artificial intelligence (AI) framework for computational science and engineering. Comput. Methods Appl. Mech. Eng. **373**, 113452 (2021)
88. T. Hastie, R. Tibshirani, J. Friedman, *The Elements of Statistical Learning* (Springer-Verlag, Heidelberg, 2002)
89. A. Agresti, *An Introduction to Categorical Data Analysis* (John Willey & Sons, Hoboken, 1996)
90. P. McCullagh, J.A. Nelder, *Generalized Linear Model* (Chapman & Hall, London, 1999)
91. J. Friedman, Multivariate adaptive regression splines. Ann. Stat. **19**(1), 1–67 (1991)
92. D. Pyle, *Data Preparation for Data Mining* (Academic Press, Waltham, 1999)
93. M. Berry, G. Linoff, *Mastering Data Mining* (John Willey & Sons, Hoboken, 2000)
94. M. Giergiel, *Modeling and Control of Mobile Robots* (PWN, Warsaw, 2002). (in Polish)

95. K. Jemielniak, *Automatic Diagnosis of Tool Condition and Cutting Process* (Warsaw University of Technology, Warsaw, 2002). (in Polish)
96. L. Rutkowski, *Methods and Techniques of Artificial Intelligence* (PWN, Warsaw, 2005). (in Polish)
97. J. Dobrodziej, A. Mazurkiewicz, J. Wojutyński, J. Ratajski, J. Michalski, Computer-aided modeling of thermal diffusion processes and PVD processes using fuzzy logic. J. Mach. Construct. Maint. **2**, 45–61 (2008). (in Polish)
98. K. Dybowski, J. Sawicki, P. Kula, B. Januszewicz, R. Atraszkiewicz, S. Lipa, The effect of the quenching method on the deformations size of gear wheels after vacuum carburizing. Arch. Metall. Mater. **61**(2), 1057–1062 (2016)
99. E. Wołowiec, P. Kula, B. Januszewicz, M. Korecki, Mathematical modelling the low-pressure nitriding process. Appl. Mech. Mater. **421**, 377–383 (2013)
100. P. Kula, E. Wołowiec, R. Pietrasik, K. Dybowski, L. Klimek, The precipitation and dissolution of alloy iron carbides in vacuum carburization processes for automotive and aircraft applications–Part I. Adv. Mater. Res. **486**, 297–302 (2012)
101. E. Wołowiec, P. Kula, The application of artificial neural networks in designing single-segment processes of vacuum carburizing. Inżynieria Materiałowa **3**, 712–715 (2010)
102. E. Wołowiec, P. Kula, The application of artificial neural networks in optimization of heat treatment processes of steel. J. Appl. Comput. Sci. **19**(1), 161–169 (2011)
103. M. Kianezhad, S. Sajjadi, H. Vafaeenezhad, A numerical approach to the prediction of hardness at different points of a heat-treated steel. J. Mater. Eng. Perform. **24**(4), 1516–1521 (2015)
104. R.G. Song, Q.Z. Zhang, Heat treatment optimization for 7175 aluminum alloy by genetic algorithm. Mater. Sci. Eng. C **17**, 133–137 (2001)
105. L. Xu, J. Xing, S. Wei, Y. Zhang, R. Long, Optimization of heat treatment technique of high-vanadium high-speed steel based on back-propagation neural networks. Mater. Design **28**(5), 1425–1432 (2007)
106. L.A. Dobrzański, W. Sitek, Application of a neural network in modelling of hardenability of constructional steels. J. Mater. Process. Technol. **78**, 59–66 (1998)
107. N. Reddy, J. Krishnaiah, S. Hong, J. Lee, Modeling medium carbon steels by using artificial neural networks. Mater. Sci. Eng. A **508**(1–2), 93–105 (2009)
108. W. Sitek, L.A. Dobrzański, J. Zacłona, The modelling of high-speed steels' properties using neural networks. J. Mater. Process. Technol. **157–158**, 245–249 (2004)
109. W. Sitek, L.A. Dobrzański, Application of genetic methods in materials' design. J. Mater. Process. Technol. **164–165**, 1607–1611 (2005)
110. T. Velsker, M. Eerme, J. Majak, Artificial neural networks and evolutionary algorithms in engineering design. J. Achieve. Mater. Manufact. Eng. **44**, 88–95 (2011)
111. A. Burghardt, T. Zaleski, Longitudinal dispersion at small and large peclet numbers in chemical flow reactors. Chem. Eng. Sci. **23**(6), 575–591 (1968)
112. O. Karabelchtchikova, *Fundamentals of Mass Transfer in Gas Carburizing*. PhD thesis (Worcester Polytechnic Institute, Worcester, 2007)
113. S. Ban-Ya, J. Elliott, J. Chipman, Thermodynamics of austenitic Fe-C alloys. Metall. Mater. Trans. A **1**(5), 1313–2130 (1970)
114. M. Bannister, Control of carbon potential using an oxygen sensor. Ind. Heat. **51**(3), 24–26 (1984)
115. O. Karabelchtchikova, S. Johnston, H. DaCosta, R. Sisson, Thermodynamic and kinetic aspects of endothermic carburizing atmospheres with natural gas enrichment, in *Proceedings of 24th Heat Treating Society Conference 2007* (Detroit, 2007), pp. 40–44
116. O. Karabelchtchikova, M. Maniruzzaman, R. Sisson, Carburization process modeling and sensitivity analysis using numerical simulation, in *Proceedings of Materials Science & Technology 2006 Conference* (Cincinnati, 2006), pp. 375–386
117. O. Karabelchtchikova, R. Sisson, Carbon diffusion in steels: A numerical analysis based on direct integration of the flux. J. Phase Equilib. Diffus. **26**(6), 598–604 (2006)

118. O. Rowan, R. Sisson, Effect of alloy composition on carburizing performance of steel. J. Phase Equilib. Diffus. **30**(3), 235–241 (2009)
119. O. Karabelchtchikova, C. Brown, R. Sisson, Effect of surface roughness on kinetics of mass transfer during gas carburising. Int. Heat Treat. Surf. Eng. **1**(4), 164–170 (2007)
120. W. Gräfen, B. Edenhofer, Acetylene low-pressure carburizing–a novel and superior carburizing technology. HTM J. Heat Treat. Mater. **26**(4), 79–83 (1999)
121. T. Kubota, *Vacuum Carburizing Method and Device, and Carburized Products*. Patent US 5,702,540 (JH Corporation, Japan, 1997)
122. R. Poor, *Vacuum Carburizing with Unsaturated Aromatic Hydrocarbons–Raport*. Patent US 7,033,446 (Surface Combustion, USA, 2006)
123. P. Kula, J. Olejnik, P. Heilman, *Hydrocarbon Gas Mixture for the Under Pressure Carburizing of Steel*. Patent EU 1,558,780 (Seco/Warwick, Lodz University of Technology, EU, 2007)
124. R. Khan, *Vacuum Gas Carburizing–fate of Hydrocarbons*. PhD thesis (Universitat Karlsruhe, Karlsruhe, 2008)
125. A. Matsugi, A. Miyoshi, Modeling of two- and three-ring aromatics formation in the pyrolysis of toluene. Proceed. Combust. Inst. **34**(1), 269–277 (2013)
126. N. Slavinskaya, U. Riedel, S. Dworkin, M. Thomson, Detailed numerical modeling of PAH formation and growth in non-premixed ethylene and ethane flames. Combust. Flame **159**(3), 979–995 (2012)
127. T. Bensabath, H. Monnier, P.-A. Glaude, Detailed kinetic modeling of the formation of toxic polycyclic aromatic hydrocarbons (PAHs) coming from pyrolysis in low-pressure gas carburizing conditions. J. Anal. Appl. Pyrol. **122**, 342–354 (2016)
128. C. Saggese, N. Sanchez, A. Frassoldati, A. Cuoci, T. Faravelli, M. Alzueta, E. Ranzi, Kinetic modeling study of polycyclic aromatic hydrocarbons and soot formation in acetylene pyrolysis. Ener. Fuels **28**(2), 1489–1501 (2014)
129. C. Xu, A. Al Shoaibi, C. Wang, H. Carstensen, A. Dean, Kinetic modeling of ethane pyrolysis at high conversion. J. Phys. Chem. A **115**(38), 10470–10490 (2011)
130. K. Norinaga, O. Deutschmann, N. Saegusa, J. Hayashi, Analysis of pyrolysis products from light hydrocarbons and kinetic modeling for growth of polycyclic aromatic hydrocarbons with detailed chemistry. J. Anal. Appl. Pyrol. **86**(1), 148–160 (2009)
131. C. Muller, V. Michel, G. Scacchi, G. Côme, THERGAS: A computer program for the evaluation of thermochemical data of molecules and free radicals in the gas phase. J. de Chimie Physique **92**, 1154–1178 (1995)
132. Detchem. DETCHEM, *Computer Application* (2008)
133. A. Smirnov, J. Kuleshov, Calculation of diluted ammonia nitriding reactions. Metallovedenie i Termicheskaya Obrabotka Metallov **5**, 45–49 (1966). (in Russian)
134. G. Ertl, M. Huber, Mechanism and kinetics of ammonia decomposition on iron. J. Catal. **61**, 537–547 (1980)
135. G. Mariadassou, C.-H. Shin, G. Bugli, Tamaru's model for ammonia decomposition over titanium oxynitride. J. Mol. Catal. A **141**, 263–267 (1999)
136. S. Yin, B. Xu, X. Zhou, C. Au, A mini-review on ammonia decomposition catalysts for on-site generation of hydrogen for fuel cell applications. Appl. Catal. A **277**, 1–9 (2004)
137. A. Waghode, N. Hanspal, I. Shigidi, V. Nassehi, K. Hellgardt, Computer modelling and numerical analysis of hydrodynamics and heat transfer in non-porous catalytic reactor for the decomposition of ammonia. Chem. Eng. Sci. **60**, 5862–5877 (2005)
138. K. Alhumaizi, R. Henda, M. Soliman, Numerical analysis of a reaction-diffusion-convection system. Comput. Chem. Eng. **27**, 579–594 (2003)
139. I. Ozdemir, F. Akar, The effect of flow orientation on nitriding process. Vacuum **116**, 104–109 (2015)
140. L. Małdziński, J. Tacikowski, New possibilities to control the gas nitriding process by using the simulation of the kinetics of the growth of nitrided layers. Inżynieria Powierzchni **2**, 3–10 (1998). (in Polish)
141. J. Michalski, J. Dobrodziej, J. Tacikowski, J. Wojutyński, A. Mazurkiewicz, J. Ratajski, P. Wach, Simulation of three-components atmospheres in control system of the gas nitriding processes. Inżynieria Materiałowa **29**(6), 848–853 (2008). (in Polish)

142. J. Michalski, T. Frączek, Mathematical models of the change in composition of the inlet atmosphere for the assumed changes in the nitrogen potential. Inżynieria Powierzchni **25**(1), 1–9 (2020). (in Polish)
143. D. Goldberg, *Genetic Algorithms* (Addison-Wesley Reading, Boston, 1989)
144. E. Vyazmina, J. Sheng, S. Jallais, L. Bustamante-Valencia, P. Bruchet, F.-P. Richard, Carbonitriding: Kinetic modeling of ammonia and acetylene decomposition at high temperature and low pressure, in *Proceeding of 24th IFHTSE Congress/European Conference on Heat Treatment, June 26–27, 2017* (Nice, 2017)
145. Ansys. CHEMKIN, *Computer Application* (2021)
146. F. Harris, Case depth–an attempt at a practical definition. Metal Progr. **44**, 265–272 (1943)
147. R. Asimow, Analysis of the variation of the diffusion constant of carbon in austenite with concentration. Trans. AIME **230**(3), 611–613 (1964)
148. J. Wunning, Advances in gas carburizing technique. HTM Härtereitechnische Mitteilungen **23**(3), 101–109 (1968)
149. K. Rimmer, E. Schwarz-Bergkampf, J. Wunning, Surface reaction rate in gas carburizing. HTM Härtereitechnische Mitteilungen **30**(3), 152–160 (1975)
150. B. Moiseev, Y. Brunzel, L. Shvartsman, Kinetics of carburizing in an endothermal atmosphere. Metal Sci. Heat Treat. **21**(6), 437–442 (1979)
151. V. Munts, A. Baskatov, Rate of carburizing of steel. Metal Sci. Heat Treat. **22**(5–6), 358–360 (1980)
152. V. Munts, A. Baskakov, Mass exchange in carburization and decarburization of steel. Metal Sci. Heat Treat. **25**(2), 98–102 (1983)
153. P. Stolar, B. Prenosil, Kinetics of transfer of carbon from carburizing and carbonitriding atmospheres. Metall. Mater. **22**(5), 348–353 (1984)
154. J. Agren, Revised expression for the diffusivity of carbon in binary Fe-C austenite. Scripta Metall. **20**(11), 1507–1510 (1986)
155. S. Gut, Evaluation of carbon transfer coefficient. Inżynieria Materiałowa **27**(5), 1002–1004 (2006)
156. S. Gut, Correction method of carbon transfer coefficient. Inżynieria Materiałowa **32**(4), 429–431 (2011). (in Polish)
157. R. Pietrasik, P. Kula, Surface phenomena accompanying the vacuum carburizing process. Inżynieria Materiałowa **5**, 437–439 (2005). (in Polish)
158. R. Gorockiewicz, A. Łapinski, Structure of the carbon layer deposited on the steel surface after low-pressure carburizing. Vacuum **85**, 429–433 (2010)
159. R. Gorockiewicz, The kinetics of low-pressure carburizing of alloy steels. Vacuum **86**(4), 448–451 (2011)
160. P. Kula, K. Dybowski, E. Wołowiec, R. Pietrasik, Boost-diffusion vacuum carburising-process optimisation. Vacuum **99**, 175–179 (2014)
161. Y. Wei, G. Wang, R. Sisson, B. Bernard, R. Poor, Intelligent heat treating: Simulation of carburizing process, in *Proceedings of the 26th ASM Heat Treating Society Conference* (Cincinnati, 2011), pp. 91–98
162. Y. Baran, E. Gyulikhandanov, High-temperature nitriding of highly dense sintered steels. Metal Sci. Heat Treat. **28**(9), 629–632 (1986)
163. H. Carslaw, J. Jaeger, *Conduction of Heat in Solids* (Oxford University Press, Glasgow, 1959)
164. W. Jost, *Diffusion in Solids, Liquids and Gases* (Academic Press, New York, 1952)
165. H. Mehrer, *Diffusion in Solids* (Methods, Materials, Diffusion-controlled Processes (Springer-Verlag, Berlin Heidelberg, Fundamentals, 2007)
166. K. Dybowski, P. Kula, Determination of the effective diffusion coefficient for the control of the vacuum carburizing process. Inżynieria Materiałowa **5**, 391–393 (2005). (in Polish)
167. K. Dybowski, R. Pietrasik, Impact of carbon deposit for the vacuum carburizing process. Inżynieria Materiałowa **5**, 939–943 (2006). (in Polish)
168. S. Tsepov, Characteristic features of carburizing of steel during vacuum carburizing. Metal Sci. Heat Treat. **21**(8), 633–638 (1979)

169. K. Bongartz, D. Lupon, H. Schuster, A model to predict carburization profiles in high temperature alloys. Metall. Trans. A **11**, 1883–1893 (1980)
170. K. Bongartz, R. Schulten, W. Quadakkers et al., A finite difference model describing carburization in high-temperature alloys. Corrosion **42**, 390–397 (1986)
171. A. Engström, L. Höglund, J. Ågren, Computer simulation of carburization in multiphase systems. Mater. Sci. Forum **163–165**, 725–730 (1994)
172. P. Cavaliere, G. Zavarise, M. Perillo, Modeling of the carburizing and nitriding processes. Comput. Mater. Sci. **46**(1), 26–35 (2009)
173. Ł. Kołodziejczyk, *Mathematical Modeling of the Process of Vacuum Carburizing*. PhD thesis (Lodz University of Technology, Lodz, 2003. (in Polish)
174. AWT-Fachausschuss 5 and Arbeitskreis 4, (ed.) *Die Prozessregelung beim Gasaufkohlen und Einsatzharten* (Expert Verlag, Renningen-Malmsheim, 1997). (in German)
175. O. Karabelchtchikova, R. Sisson, Calculation of gas carburizing kinetics from carbon concentration profiles based on direct flux integration. Defect Diffus. Forum **266**, 171–180 (2007)
176. L. Zhang, *Modeling and Verification of Simulation Tools for Carburizing and Carbonitriding*. PhD thesis (Worcester Polytechnic Institute, Worcester, 2016)
177. S. Wei, G. Wang, X. Zhao, X. Zhang, Y. Rong, Experimental study on vacuum carburizing process for low-carbon alloy steel. J. Mater. Eng. Perform. **23**(2), 545–550 (2014)
178. R. Smith, The diffusivity of carbon in Fe-Co alloys. Trans. Metall. Soc. AIME **230**, 470–480 (1964)
179. M. Krishtal, Diffusion processes in iron alloys. Jerusalem **1963**, 90–133 (1970)
180. M. Ismail, S. Iskander, E. Saleh, Carburizing of steels. Surf. Technol. **12**(4), 341–349 (1981)
181. X. Zhang, R. Takahashi, T. Akiyama, J. Yagi, Carburization rate into solid iron at CO-CO2 atmosphere. Tetsu-to-Hagane **83**(5), 299–304 (1997)
182. T. Murakami, H. Fukuyama, K. Nagata, Mechanisms of carburization and melting of iron by CO gas. ISIJ Int. **41**, 416–421 (2001)
183. K. Dybowski, *The Computing of Effective Carbon Diffusion Coefficient in Steels to a Process of Vacuum Carburizing Control*. PhD thesis (Lodz University of Technology, Lodz, 2005). (in Polish)
184. S.-J. Lee, D. Matlock, C. van Tyne, An empirical model for carbon diffusion in austenite incorporating alloying element effect. ISIJ Int. **51**, 1903–1911 (2011)
185. M. Zajusz, K. Tkacz-Śmiech, M. Danielewski, Modeling of vacuum pulse carburizing of steel. Surf. Coat. Technol. **258**, 646–651 (2014)
186. D. Herring, Simulation software (Part one: Atmosphere carburizing). Ind. Heat. **1**, 14–16 (2018)
187. R. Alquicer, Controls software in atmosphere equipment. Gear Solutions **1**, 26–27 (2016)
188. D. Herring, *Atmosphere Heat Treatment: Principles, Applications, Equipment*, vol. I (BNP Media Group, Birmingham, 2014)
189. K. Loeser, B. Gornicki, Low pressure carburizing of large transmission parts. Gear Technol. **9–10**, 67–70 (2009)
190. O. Karabelchtchikova, G. Wang, R. Sisson, New carburizing calculation tool for gas and low-pressure carburizing. Heat Treat. Progr. **3–4**, 18 (2008)
191. Y. Wei, *Simulation, Optimization and Development of Thermo-chemical Diffusion Processes*. PhD thesis (Worcester Polytechnic Institute, Worcester, 2013)
192. M. Jaster, Why vacuum carburizing? Gear Solutions **3–4**, 31–35 (2010)
193. D. Herring, *Vacuum Heat Treatment: Principles, Practices, Applications* (BNP Media II, Troy, 2012)
194. F. Otto, D. Herring, Vacuum carburizing of aerospace and automotive materials. Heat Treat. Progr. **1–2**, 33–37 (2005)
195. M. Lohrmann, W. Gräfen, D. Herring, J. Greene, Acetylene vacuum carburising (AvaC) as the key to the integration of the case-hardening process into the production line. Heat Treat. Metals **29**(2), 39–43 (2002)
196. P. Kula, R. Pietrasik, E. Wołowiec, B. Januszewicz, A. Rzepkowski, Low-pressure nitriding according to the FineLPN technology in multi-purpose vacuum furnaces. Adv. Mater. Res. **586**, 230–234 (2012)

197. P. Kula, J. Olejnik, J. Kowalewski, Smart control system optimizes vacuum carburizing process. Ind. Heat. **9**, 99–102 (2003)
198. P. Kula, J. Olejnik, J. Kowalewski, Smart system for vacuum carburizing. Aluminium **16**(3), 3–5 (2004)
199. M. Górecki, P. Kula, Computer aided simulation of the vaccum carburizing process of elements containing deep holes. Inżynieria Materiałowa **5**, 591–596 (2004)
200. P. Kula, J. Olejnik, P. Heilman, *Method for Under-pressure Carburizing of Steel Workpieces*. Patent EU 1,558,781 (Seco/Warwick, Lodz University of Technology, EU, 2006)
201. Lodz University of Technology. *SimVac. Utility Model Z-352663* (Lodz University of Technology, Lodz, 2009)
202. P. Kula, Ł. Kołodziejczyk, D. Siniarski, A. Krasiński, M. Górecki, The use of a numerical model in the design of vacuum carburizing. Inżynieria Materiałowa **4**, 511–513 (2001). (in Polish)
203. E. Wołowiec, P. Kula, Ł. Kołodziejczyk, K. Dybowski, M. Korecki, Mathematical modelling of the vacuum carburizing process. Ther. Process. Gear Solutions **3–4**, 34–40 (2014)
204. P. Kula, M. Korecki, R. Pietrasik, E. Wołowiec, K. Dybowski, Ł. Kołodziejczyk, R. Atraszkiewicz, M. Krasowski, FineCarb—the flexible system for low-pressure carburizing. New options and performance. Japan Soc. Heat Treat. **49**, 133–136 (2009)
205. S. Hiramatsu, K. Inagaki, H. Sakaue, I. Yamamoto, Study of the high-temperature vacuum carburizing phenomenon and the carburizing prediction model based by big data approach, in *Proceeding of 24th IFHTSE Congress/European Conference on Heat Treatment, June 26–27, 2017* (Nice, 2017)
206. K. Genel, I. Ozbek, A. Kurt, C. Bindal, Boriding response of AISI W1 steel and use of artificial neural network for prediction of borided layer properties. Surf. Coat. Technol. **160**, 38–43 (2002)
207. E. Wołowiec-Korecka, Methods of data mining for modelling of low-pressure heat treatment. J. Ach. Mater. Manuf. Eng. **1**(85), 31–40 (2017)
208. J. Tacikowski, J. Zyśk, *Method of Gas Nitriding*. Patent PL 85924 (Institute of Precision Mechanics, Poland, 1977)
209. L. Małdziński, J. Tacikowski, Concept of an economical and ecological process of gas nitriding of steel. HTM Hartereitechnische Mitteilungen **61**(6), 296–302 (2006)
210. L. Małdziński, Controlled nitriding using a ZeroFlow process. Heat Treat. Progr. **8**, 53–55 (2007)
211. L. Małdziński, M. Bazel, M. Korecki, A. Miliszewski, T. Przygoński, Industrial experiences with controlled nitriding using a ZeroFlow method. Heat Treat. Progr. **7–8**, 19–22 (2009)
212. Y. Lakhtin, *Physical Foundations of the Nitriding Process* (Mashgiz, Moscow, 1948). (in Russian)
213. Y. Lakhtin, Diffusion foundations of the nitriding process. Metal Sci. Heat Treat. **37**(7–8), 276–279 (1995)
214. M. Krukovich, Simulation of the nitriding process. Metal Sci. Heat Treat. **46**(1–2), 25–31 (2004)
215. J. Ratajski, *Selected Aspects of Modern Gas Nitriding in Terms of Process Control* (Koszalin University of Technology, Koszalin, 2003). (in Polish)
216. J. Ratajski, *Mathematical Modeling of the Gas Nitriding Process* (Koszalin University of Technology, Koszalin, 2011). (in Poland)
217. J. Ratajski, R. Olik, T. Suszko, J. Dobrodziej, J. Michalski, Design, control and in situ visualization of gas nitriding processes. Sensors **10**(1), 218–240 (2010)
218. T. Liapina, *Phase Transformations in Interstitial Fe-N Alloys*. PhD thesis (Universitat Stuttgart, Stuttgart, 2005)
219. U. Afzaal, *Modeling of Gas Nitriding using Artificial Neural Networks*. Master's thesis (McMaster University, Hamilton, 2006)
220. D. Lipiński, J. Ratajski, Modeling of microhardness profile in nitriding processes using artificial neural network. Lect. Notes Artif. Intell. **4682**, 245–252 (2007)

221. G. Khalaj, Artificial neural network to predict the effects of coating parameters on layer thickness of chromium carbonitride coating on pre-nitrided steels. Neural Comput. Appl. **23**(3–4), 779–786 (2013)
222. M. Kosikowski, Z. Suszyński, R. Olik, J. Ratajski, T. Suszko, The application of artificial neural networks and evolutionary algorithm for the designing of gas nitriding process. Intell. Inf. Eng. Syst. **13**, 33–39 (2009)
223. J. Dobrodziej, A. Mazurkiewicz, J. Wojutyński, J. Michalski, J. Tacikowski, J. Ratajski, Use of fuzzy logic to computer aid of projecting thermo-chemical processes. Inżynieria Materiałowa **29**(6), 843–847 (2008). (in Polish)
224. J. Kowalska, L. Małdziński, ZeroFlow-new, environmentally friendly method of controlled gas nitriding used for selected car parts. IOP Conf. Ser. Mater. Sci. Eng. **148**, 012047 (2016)
225. L. Małdziński, *Thermodynamic, Kinetic and Technological Aspects of Producing Nitrided Layers on Iron and Steel in Processes of Gas Nitriding* (Poznan University of Technology, Poznan, 2002). (in Polish)
226. P. Kula, *Surface Layer Engineering* (Lodz University of Technology, Lodz, 2000). (in Polish)
227. J. Michalski, *Characteristics and Calculations Atmospheres for Controlled Gas Nitriding of Steel* (Institute of Precision Mechanics, Warsaw, 2010). (in Polish)
228. J. Dobrodziej, J. Wojutyński, K. Matecki, A. Gospodarczyk, J. Michalski, J. Tacikowski, P. Wach, J. Ratajski, R. Olik, The possibility of use of computer applications to design, simulation and verification of the processes of regulated gas nitriding. Inżynieria Powierzchni **2**, 34–45 (2009). (in Polish)
229. M. Yang, R. Sisson, Alloy effects on the gas nitriding process. J. Mater. Eng. Perform. **23**(12), 4181–4186 (2014)
230. W. Dowling, T. Pattok, B. Ferguson, Development of a carburizing and quenching simulation tool: Program overview. HTM J. Heat Treat. Mater. **1**, 1–6 (1997)
231. Z. Li, R. Grandhi, R. Shivpuri, Optimum design of the heat-transfer coefficient during gas quenching using the response surface method. Int. J. Mach. Tools Manuf. **42**(5), 549–558 (2002)
232. D. Watt, L. Coon, M. Bibby, J. Goldak, C. Henwood, An algorithm for modeling of micro structural development in weld heat affected zones (Part A) reaction kinetics. Acta Metall. **36**, 3029–3035 (1988)
233. R. Wallis, N. Bhathena, P. Bhowal, E. Raymond, Application of process modeling to heat treatment of superalloys. Ind. Heat. **55**, 525–534 (1988)
234. R.A. Wallis, P.R. Bhowal, N.M. Bhathena, E.L. Raymond, Modeling the heat treatment of superalloy forgings. JOM **41**(2), 35–37 (1989)
235. J. Wang, J. Gu, X. Shan, X. Hao, N. Chen, W. Zhang, Numerical simulation of high pressure gas quenching of H13 steel. J. Mater. Process. Technol. **202**(1–3), 188–194 (2008)
236. R. Kopun, L. Škerget, M. Hriberšek, D. Zhang, B. Stauder, D. Greif, Numerical simulation of immersion quenching process for cast aluminium part at different pool temperatures. Appl. Therm. Eng. **65**(1–2), 74–84 (2014)
237. A.E.A. Technology, *CFX-5 Solver and Solver Manager* (AEA Technology, CFDS Department, 2002)
238. J. Mackerle, Finite element analysis and simulation of quenching and other heat treatment processes. Comput. Mater. Sci. **27**(3), 313–332 (2003)
239. F. Krause, S. Schüttenberg, U. Fritsching, Modelling and simulation of flow boiling heat transfer. Int. J. Numer. Methods Heat Fluid Flow **20**, 312–331 (2010)
240. P.K. Meduri, G.R. Warrier, V.K. Dhir, Wall heat flux partitioning during subcooled forced flow film boiling of water on a vertical surface. Int. J. Heat Mass Transf. **52**, 3534–3546 (2009)
241. V. Srinivasan, D.M. Wang, Modelling and simulation of the heat and mass transfer characteristics of binary mixtures for boiling flow applications. J. ASTM Int. **8**, 1–22 (2011)
242. B. Gao, H. Li, Y. Chen, J. Dong, Numerical analysis and structure improvement for the corrosion and cracking of the mixing tee for the quenching gas of hydrogen from coal. Procedia Eng. **130**, 1246–1257 (2015)

243. K. Krupanek, J. Sawicki, V. Buzalski, Numerical simulation of phase transformation during gas quenching after low pressure carburizing. IOP Conf. Ser. Mater. Sci. Eng. **743**(1), 012047 (2020)
244. Z. Gawroński, J. Sawicki, Toothed wheel optimization by means of the finite element analysis. Mech. Mech. Eng. **4**(2), 183–189 (2000)
245. J. Sawicki, P. Siedlaczek, A. Staszczyk, Finite-element analysis of residual stresses generated under nitriding process: A three-dimensional model. Metal Sci. Heat Treat. **59**(11–12), 799–804 (2018)
246. L. Huiping, Z. Guoqun, N. Shanting, L. Yiguo, Optimization of technology parameters for the plane-strain component in the process of gas quenching. Appl. Math. Modell. **32**(5), 860–872 (2008)
247. D. Herring, A review of gas quenching from the perspective of the heat transfer coefficient. Ind. Heat. **2**, 67–72 (2006)
248. B.H. Morales, J.K. Brimacombe, E.B. Hawbolt, Determination of quench heat transfer coefficients using inverse techniques, in *Proceedings of the 1st International Conference on Quenching and Control of Distortion* (Chicago, 1996), pp. 155–164
249. J. Davis, G. Davidson, S. Lampman, T. Zorc, J. Daquila, A.S.M. Handbook, *Heat Treating*, vol. 4 (ASM, Materials Park, 2011)
250. Y.H. Guan, T.L. Chen, G.H. Wang, J.T. Zhang, The prediction of the mechanical properties of metal during laser quenching. J. Mater. Process. Technol. **63**(1–3), 614–617 (1997)
251. S. Denis, D. Farias, A. Simon, Mathematical-model coupling phase-transformations and temperature evolutions in steels. ISIJ Int. **32**, 316–325 (1992)
252. C. Magee, Phase transformations, in *Proceedings of American Society for Metals Seminar* (Metals Park, 1970), pp. 115–154
253. B. Podgornik, I. Belič, V. Leskovšek, M. Godec, Tool steel heat treatment optimization using neural network modeling. Metall. Mater. Trans. A **47**(11), 5650–5659 (2016)
254. G. Cybenko, *Continuous Valued Neural Networks with Two Hidden Layers are Sufficient* (Technical report (Tufts University, Medford, 1988)
255. H. Cheng, H. Wang, J. Xie, Calculation of coupled problem between temperature and phase transformation during gas quenching in high pressure. Appl. Math. Mech. **27**(3), 305–311 (2006)
256. P. Ferro, H. Porzner, A. Tiziani, F. Bonollo, The influence of phase transformations on residual stresses induced by the welding process–3D and 2D numerical models. Modell. Simul. Mater. Sci. Eng. **14**(2), 117–136 (2006)
257. M. Dalgic, G. Löwisch, Transformation plasticity at different phase transformations of bearing steel. Materialwissenschaft und Werkstofftechnik **37**(1), 122–127 (2006)
258. J. Leblond, G. Mottet, J. Devaux, A theoretical and numerical approach to the plastic behaviour of steels during phase transformations–I. Derivation of general relations. J. Mech. Phys. Solids **34**, 395–409 (1986)
259. J.B. Leblond, G. Mottet, J.C. Devaux, A theoretical and numerical approach to the plastic behaviour of steels during phase transformations–II. Study of calssical plasticity for ideal-plastic phases. J. Mech. Phys. Solids **34**, 411–432 (1986)
260. J.B. Leblond, J. Devaux, J.C. Devaux, Mathematical modelling of transformation plasticity in steels I: Case of ideal-plastic phases. Int. J. Plast. **5**, 551–572 (1989)
261. E. Mittemeijer, M. Somers (eds.), *Thermochemical Surface Engineering of Steels* (Woodhead Publishing, Cambridge, 2015)
262. R. Atraszkiewicz, B. Januszewicz, Ł Kaczmarek, W. Stachurski, K. Dybowski, A. Rzepkowski, High pressure gas quenching: Distortion analysis in gears after heat treatment. Mater. Sci. Eng. A **558**, 550–557 (2012)
263. H. Li, G. Zhao, C. Huang, S. Niu, Technological parameters evaluation of gas quenching based on the finite element method. Comput. Mater. Sci. **40**, 282–291 (2007)
264. J. Sawicki, B. Kruszyński, R. Wójcik, The influence of grinding conditions on the distribution of residual stress in the surface layer of 17CrNi6-6 steel after carburizing. Adv. Sci. Technol. Res. J. **11**(2), 17–22 (2017)

265. T. Inoue, D. Ju, K. Arimoto, Metallo-thermo-mechanical simulation of quenching process—theory and implementation of computer code HEARTS, in *Proceedings of the 1st International Conference on Quenching and Control of Distortion* (Chicago, 1992), pp. 205–212
266. B. Ferguson, Effective technical collaboration in heat treatment process modelling: A case study. Int. Heat Treat. Surf. Eng. **6**(2), 61–66 (2012)
267. Z. Li, B. Ferguson, V. Nemkov, R. Goldstein, J. Jackowski, G. Fett, Effect of quenching rate on distortion and residual stresses during induction hardening of a full-float truck axle shaft. J. Mater. Eng. Perform. **23**(12), 4170–4180 (2014)
268. E. Nart, K. Genel, Investigation of carburising treatment of injector flange in diesel engine. Mater. Design **31**(4), 2110–2117 (2010)
269. H. Adrian, Thermodynamic model for precipitation of carbonitrides in high strength low alloy steels containing up to three microalloying elements with or without additions of aluminium. Mater. Sci. Technol. **8**(5), 406–420 (1992)
270. H. Adrian, *Numerical Modeling of Heat Treatment Processes* (AGH University, Cracow, 2011). (in Polish)
271. P. Marynowski, H. Adrian, M. Głowacki, Modeling of the kinetics of carbonitride precipitation process in high-strength low-alloy steels using cellular automata method. J. Mater. Eng. Perform. **28**(7), 4018–4025 (2019)
272. F. Szidarowszky, S. Yakowitz, *Principles and Procedures of Numerical Analysis* (Plenum Press, New York, 1978)
273. S. Sahay, K. Mitra, Cost model based optimisation of carburising operation. Surf. Eng. **20**(5), 379–384 (2004)
274. H. Antes, Calculating the gas flow rate for vacuum carburization. Heat Treat. Progr. **8**, 51–53 (2005)
275. E. Wołowiec-Korecka, M. Korecki, M. Sut, A. Brewka, P. Kula, Calculation of the mixture flow in a low-pressure carburizing process. Metals **9**(4), 439–446 (2019)
276. H. Ikehata, K. Tanaka, H. Takamiya, H. Mizuno, T. Shimada, Modeling growth and dissolution kinetics of grain-boundary cementite in cyclic carburizing. Metall. Mater. Trans. A **44**(8), 3484–3493 (2013)
277. A. Schneider, Iron layer formation during cementite decomposition in carburising atmosperes. Corros. Sci. **44**, 2353–2365 (2002)
278. P. Saltykov, M. Putilina, Three effects of the cementite phase of iron-carbon alloys on their electrochemical behaviour in an axalic acid medium. Protect. Metals Phys. Chem. Surf. **42**, 372–399 (2006)
279. S. Yusa, T. Hara, K. Tsuzaki, Refinement of grain boundary cementite in medium-carbon tempered martensite by thermomechanical processing. Mater. Sci. Eng. A **273–275**, 462–465 (1999)
280. E. Głowacz, H. Adrian, W. Osuch, The nitrogen content effect of carbonitride coagulation in 40Cr8 steel with micro-additions V and V+Al. Arch. Metall. Mater. **58**(2), 607–611 (2013)
281. P. Kula, E. Wołowiec, The application of artificial intelligence to modelling and evaluation of machines parts, in *Applications of Systems Science*. ed. by A. Grzech, P. Świątek, K. Brzostowski (Warszawa, EXIT, 2010), pp. 315–320
282. E. Wołowiec, K. Dybowski, R. Atraszkiewicz, SimVaC–Advanced capabilities of the FineCarb vacuum carburizing software, in *Prceedings of 12th Seco/Warwick Seminar, September 25–26, 2008* (Bukowy Dworek, 2008), pp. 27–33. (in Polish)
283. E. Wołowiec, P. Kula, M. Korecki, J. Olejnik, Simulation and control of tool steel quenching process, in *Proceeding of 25th European Conference on Modelling and Simulation, June 7–10, 2011* (Cracow, 2011), pp. 357–361
284. M. Kuhn, K. Johnson, *Applied Predictive Modelling* (Springer, New York, 2013)
285. S. Skansi, *Introduction to Deep Learning: From Logical Calculus to Artificial Intelligence* (Springer, Cham, 2018)
286. F. Collet, *Deep Learning with Python* (Manning Publications Co., Shelter Island, 2021)

Chapter 7
Looking at Future Challenges

The resources of civilization are not yet exhausted.

Wiliam Gladstone, 1809–1898

According to the Industrial Heating Equipment Association (IHEA), heat and thermo-chemical treatment accounts for 2–15% of total production costs. According to Grand View Research, in 2021 the value of the global service sector for these treatments was US$100 billion, of which approximately 15% was vacuum treatments [1–3]. The main customers for these services are the automotive, heavy construction machinery, aerospace, mining, metalworking and tool and die industries (Fig. 7.1) and demand for heat treatment services and equipment continues to grow [2]. Currently, the following needs are the greatest driving force behind the development of new heat treatment technologies and methods, which partly coincide with expectations signalled two decades earlier (see Chap. 2), at the dawn of the twenty-first century:

1. shortening the process time,
2. minimising deformations of manufactured parts,
3. reducing the environmental footprint related to the operation of heat treatment equipment;
4. reducing the overall cost of heat treatment, including the operation of treatment.

There is also a noticeable environmental factor, influenced by the growing awareness of global climate risks.

Change in production philosophy

A gradual change in production philosophy is observed when analysing the scientific and industrial literature of the last two decades. The world is gradually

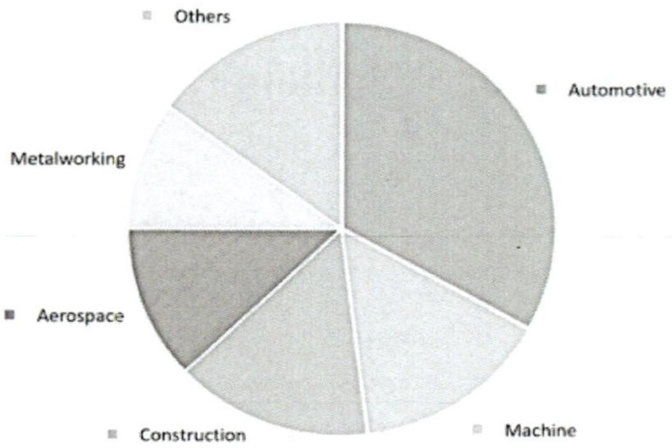

Fig. 7.1 Share of industrial segments in the global market of heat and thermo-chemical treatment services. Based on data of grand view research in Ref. [3]

shifting from mass to high volume production, reducing the quantity of products and increasing their quality. As regards the evolution in the construction of automotive vehicles, fewer power transmission components are observed in the construction of electric cars, but at the same time, the new transmission mechanisms are more precisely made, resulting in reduced noise, reduced resistance and thermal losses and improved energy efficiency. In the end, this translates into a reduction of costs (see Chap. 5); with the same market consumption we produce less and at the same time more accurately—fewer production rejects are generated, costs and finishing treatment times are reduced. We are creating more consciously, satisfying our needs and at the same time trying not to overburden the environment with human activity. The growing number of scientific publications, industry reports and EU directives from year to year suggests that this trend will continue in the coming decades.

Increasing share of vacuum heat treatment

A clear trend currently occurring in heat and thermo-chemical treatment is the heightening use of vacuum treatment, in particular low-pressure carburising (LPC). The largest use of vacuum equipment is in the aerospace, automotive, medical, tool and die industries. In 2016, it was predicted that amongst the listed industries, the share of the automotive industry would increase significantly, due to the growing acceptance of low-pressure carburising for carburising automotive transmission components. Today, LPC is one of the fastest growing heat treatment processes, especially in the sector manufacturing gearboxes for vehicles, and it is estimated that up to 40% of powerdrive components are carburised in vacuum furnaces using LPC technology, while traditional atmospheric carburising furnaces still dominate heat treatment in ship, truck and heavy machinery manufacturing [2].[1]

Increasing digitalisation and standardisation of heat treatment operations

A dynamic increase in the application of computer-based computational methods in heat treatment in scientific aspects is seen by analysing the literature published between 2001 and 2022. This phenomenon is also occurring in the industrial field and is visible as a growing trend of integrating heat treatment into the production process, especially increasing the level of controllability and ease of maintenance of the device through the use of the "Internet of Things" concept and systems integration (see Chap. 6). Over the past 15 years, machining design systems and control systems for machining equipment have changed dramatically. Nowadays, there is no doubt that control systems speed up planning and machining processes, minimise erroneous decisions and events, make the process smoother and generally reduce the production costs. In addition, it is believed that there will be a shift from mass production to a model of serial production and one-piece production in the future, as a result of increasing customer demands for customised products and the ability to trace production flows at low costs (see Chaps. 3 and 4). This trend will only boost the importance of computer-based design and supervision systems. The phenomenon will spread with the popularisation of the ideas of Industry 4.0 and Society 5.0. Digitalisation is followed by the globalisation of heat treatment standards, including procedures, processes and equipment design. The importance of safety in the operation of heat treatment equipment is growing, thanks to the philosophy of designing accident-free systems and equipment, and the trend towards the concept of management by quality is intensifying, which is possible primarily through the standardisation of processes and procedures (see Chap. 1).

Increasing influence of additive manufacturing methods

Additive manufacturing methods (3D printing) are one of the pillars of the concept of H. Kagermann et al. (Industry 4.0). Although it is still too early to say clearly how these methods will find their permanent place amongst manufacturing methods in the next two decades, they are currently being intensively developed and researched. It is anticipated that they will have an impact on heat treatment methods for metal components. For the efficient operation of such products, it is necessary to develop

[1] By 2015, more than 1500 LPC chambers have been installed around the world. Volkswagen, GM, BMW, Ford, CFA, Renault, Porsche, Honda, Hyundai and Indian car manufacturers (although with some limits) such as Mahindra and Tata have vacuum Low-Pressure Carburizing systems for heat treatment its car gearboxes. The majority of gear manufacturers like EATON, ZF, GETRAG (MAGNA), Hoerbiger, JATCO, AISIN SEIKI, AICHI KIKAI, SAGW, GREAT WALL, GKN, LINAMAR and DANA use LPC processes for heat treatment gears. The new generation of transmissions such as the dual clutch (DCT) and six-eight-nine speed automated transmissions are all carburised with LPC technology [4].

dedicated thermal improvement technologies and, in particular, equipment to carry out these treatments.

Increasing ecological care and reducing the carbon footprint

Practically since the beginning of the twenty-first century, a growing trend towards attention to the environment and other ecological aspects has been observed. This is linked to the Third Industrial Revolution (Industry 3.0), which meant that places for industrial development and human settlements were no longer dependent on the distribution of raw materials and energy sources, but on a clean, human-friendly environment. This trend is resulting in modifications to heat treatment technologies with a view to reducing factors that are harmful to the environment and humans. The next few years should bring modifications or even replacement of traditional process atmospheres with low-pressure atmospheres and quenching agents, particularly quenching oils, with neutral gases.

As A. Bhaskar aptly put it, steel is the backbone of modern civilisation—nearly 2 billion tonnes of this material were produced worldwide in 2021 [5, 6]. At the same time, metallurgy, together with iron and steel making, is one of the most energy-intensive manufacturing sectors, using up to 5% of the world's total energy consumption and accounting for a third of global industrial CO_2 emissions. This puts pressure on the industry to evolve towards more sustainable ways of production [7, 8]. In 2016, the iron and steel industry in the EU-28 was responsible for releasing 200 million tonnes of CO_2 into the atmosphere, meanwhile, according to the EU Roadmap 2050, iron and steel emissions are expected to be reduced by around 85% [9–11]. This signifies a shift towards a complete change in steel production technology in Europe in the coming years.

It seems obvious that a green revolution encompassing steel production will not bypass the area of heat and thermo-chemical treatment either, especially where a zero carbon footprint will be declared. There is no doubt that moving away from traditional processes will require extensive research and development to resolve the issues and barriers facing new, breakthrough technologies.

References

1. Industrial Heating Equipment Association. The Industrial Heating Equipment Association (IHEA) official page (2016)
2. J. Kowalewski, Ten new trends in heat treatment and quenching: A global perspective, in *Proceedings of 3rd International Conference on Heat Treatment and Surface Engineering in Automotive Applications, May 11–13, 2016* (Prague, 2016)
3. Grand View Research. Heat treating market size, share & trends analysis report by material (steel, cast iron), by process (case hardening, annealing), by equipment, by application, by region, and segment forecasts, 2022-2030. Report GVR-2-68038-287-7 (Grand View Research, San Francisco, 2021)
4. J. Kowalewski, *Amazing Vacuum Furnaces–LPC Technology Within Gear Industry* (2016)

5. A. Bhaskar, M. Assadi, H. Nikpey Somehsaraei, Decarbonization of the iron and steel industry with direct reduction of iron ore with green hydrogen. Energies **13**(3), 758 (2020)
6. A. de la Garza, Green steel from Sweden. Time Mag. **199**(17–18), 14–15 (2022)
7. M. Quader, S. Ahmed, R. Ghazilla, S. Ahmed, M. Dahari, A comprehensive review on energy efficient CO2 breakthrough technologies for sustainable green iron and steel manufacturing. Renew. Sustain. Energy Rev. **50**, 594–614 (2015)
8. E. Karakaya, C. Nuur, L. Assbring, Potential transitions in the iron and steel industry in Sweden: Towards a hydrogen-based future? J. Clean. Prod. **195**, 651–663 (2018)
9. B. Borkent, J. de Beer, *Carbon Costs for the Steel Sector in Europe Oost-2020-impact Assessment of the Proposed ETS Revision* (Technical report, ECOFYS, Ultrecht, 2016)
10. M. Anderl, J. Burgstaller, B. Gugele, *Klimaschutzbericht 2018*. Technical Report REP-0660 (Umweltbundesamt GmbH, Wien, 2018). (in German)
11. M. Hammerschmid, S. Müller, J. Fuchs, H. Hofbauer, Evaluation of biomass-based production of below zero emission reducing gas for the iron and steel industry. Biomass Convers. Biorefin. **11**(1), 169–187 (2021)

Chapter 8
Strategic Directions in the Development of Global Heat Treatment

The global directions for improving heat and thermo-chemical treatment processes and materials in 2000–2020 included a few crucial aspects. Three of them are presented in the table below; all of them are described in detail in the works cited at the foot of the table. The term *Profit* means an expected financial return for the heat treatment division (excluding environmental initiatives); benefits from environmental initiatives include environmental value and public perception of the heat treatment industry. A high rate of return is financially significant for heat treatment enterprises and justifies the costs of their implementation in industrial practice (Table 8.1).

Table 8.1 Global directions in activities related to the improvement of heat and thermo-chemical treatment processes and materials in 2000–2020

Lp	Activity	Priority	Profit
Knowledge about materials and processes			
1	Development of improved tools for designing induction coils, reducing the number of trials and errors in pilot processes	Highest	High
2	Development of new grades of steels for high-temperature carburizing that are resistant to grain growth $>1010\,^{\circ}C$	High	High
3	Study of the feasibility and implications of high-temperature carburizing ($>1010\,^{\circ}C$), including atmospheric and vacuum carburizing processes, for shortening the process	High	High
4	Development of shorter nitriding processes	High	High
5	Determination of the kinetics of phase transformations during rapid heating for the reduction of heating time of components for heat treatment	High	High
6	Development of materials and alternative processes to carburizing, e.g. steels with limited hardenability and modified hardening processes, e.g. intensive hardening	Medium	High

(continued)

Table 8.1 (continued)

Lp	Activity	Priority	Profit
7	Development of alternative surface hardening methods for high-volume and mass production	Medium	High
8	Improving the understanding of tempering/ageing processes to decrease the variability and duration of inductive, magnetic processes	Medium	High
9	Improving the efficiency of surface modification processes	Medium	Low
10	Determination of the effect of rare earth elements on microstructural changes during heat treatment	Medium	Medium
11	Development of materials for sub-zero treatment and improving the understanding of the limitations of this treatment	Low	Low
12	Development of heat treatment processes for new materials, including ferrous and nonferrous alloys, composites and polymers	Low	Low
Knowledge of application of materials: databases or software and models			
1	Development of software for the selection of materials comprising a database of material properties where the user enters the desired properties and the results are proposals of materials and their heat treatment requirements	Low	Low
2	Development of models of processes that can relate material characteristics to operational wear/destruction, where the user enters the criteria specific to the application (e.g. wear, forces, corrosion) and the result is a specification of the material/microstructure	Low	Low
Development of computational models			
1	Development of databases concerning phase transformations under continuous heating and continuous cooling for a wide range of materials, including data concerning the effects of variations in chemical composition and heterogeneity of microstructure on transformation kinetics	Highest	High
2	Development of databases of thermal and mechanical properties in the temperature range from room temperature to heat treatment temperature	Highest	High
3	Development of data correlating heat treatment with final properties	Highest	High
4	Development of a model for geometric deformations (volumetric changes) resulting from transformations occurring during heating and cooling	High	High
5	Integration of the model of phase transformations with computer software (computer tools available on the market)	High	High
6	Development of economical methods of acquiring data from the database	High	High
7	Development of measurement methods for determining heat transfer characteristics in production equipment for applications with different media and components	High	High

Table 8.1 (continued)

Lp	Activity	Priority	Profit
8	Development of industry standards for testing & obtaining data for transformation models, including heating in furnaces and other media; quenching in oil, water, polymers, salt, air & and vacuum; spray cooling	High	High
9	Development of a thermomechanical model considering material-atmosphere interactions	High	High
10	Combining the models with process control systems for real-time process control	High	High
11	Development of thermo-chemical models of atmosphere-material interactions during carburizing, nitriding and high-temperature carburizing processes	Medium	High
12	Development of a thermomechanical model for predicting cooling rates, residual stresses and properties of machine parts	Medium	High
13	Further development of numerical fluid mechanics models to understand and achieve more efficient, homogeneous heat flow	Medium	Medium

Source R. Gaster, D. Apelian. Global database for heat treating R&D. Heat Treating Progress 9–10 (2002) 47–49; R. Gaster. Heat Treating Technology Roadmap update 2004, part I: Introduction, Heat Treating Progress 1–2 (2004) 66–67; W. Titus. Heat Treating Technology Roadmap update 2004, part II: Equipment & hardware materials. Heat Treating Progress 3–4 (2004) 39–41; R. Houghton. Heat Treating Technology Roadmap update 2004, part II: Process & materials technology. Heat Treating Progress 5–6 (2004) 45–48; S. Sikirica, D. Welling. Heat Treating Technology Update 2004, part IV: Energy & environment technology. Heat Treating Progress 7–8 (2004) 47–52; D. Apelian, R. Gaster. Heat Treating Technology Update 2004, part V: Institutional initiatives. Heat Treating Progress 9–10 (2004) 30–35; R. Houghton. Heat Treating Technology Roadmap update 2006. Part I: Process & materials technology. Heat Treating Progress 5–6 (2006) 1–4; S. Sikirica, D. Welling. Heat Treating Technology Roadmap update 2006. Part II: Energy and environmental technology. Heat Treating Progress 8–9 (2006) 1–4; G. Pfaffmann. Heat Treating Technology Roadmap update 2006. Part III: Equipment and hardware materials technology. Heat Treating Progress 11–12 (2006) 11–12; ASM. Heat Treating Technology Roadmap update 2007. Vision 2020: Looking ahead. Heat Treating Progress 1–2 (2007) 1–2

Index